Introduction to Cosmology

Plate 1

Reproduced by permission of Microcosm, CERN

Plate 2

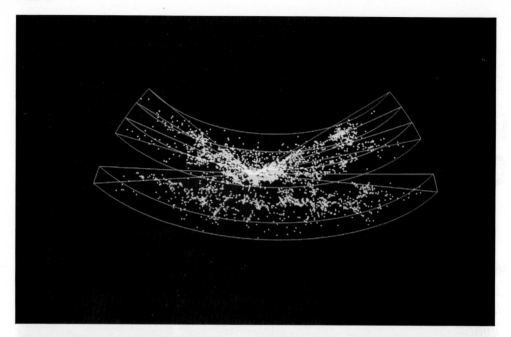

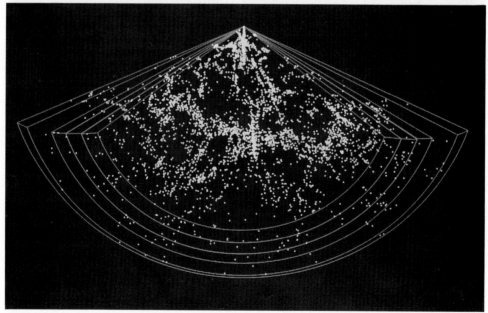

Fig. 1 Two views of the distribution of 3962 galaxies in three-dimensional redshift space. The white points display the distribution of galaxies in the right ascension range 8^h to 17^h and declination range $26.5°$ to $44.5°$; the orange in the declination range $8.5°$ to $14.5°$. 'The Great Wall' crosses the survey nearly parallel to the outer boundary. Graphics by M.J Geller, J.P. Huchra, E. Falco and R.K. McMahan, reproduced with permission from Geller *et al.*, *Science*, **246**, cover. Copyright 1989 by the AAAS

Introduction to Cosmology

Matts Roos
Laboratory of High Energy Physics
University of Helsinki
Finland

JOHN WILEY & SONS
Chichester · New York · Brisbane · Toronto · Singapore

Other Wiley Editorial Offices

John Wiley & Sons, Inc., 605 Third Avenue,
New York, NY 10158-0012, USA

Jacaranda Wiley Ltd, 33 Park Road, Milton,
Queensland 4064, Australia

John Wiley & Sons (Canada) Ltd, 22 Worcester Road,
Rexdale, Ontario M9W 1L1, Canada

John Wiley & Sons (SEA) Pte Ltd, 37 Jalan Pemimpin #05-04,
Block B, Union Industrial Building, Singapore 2057

Library of Congress Cataloging-in-Publication Data

Roos, Matts.
    Introduction to cosmology   /   Matts Roos.
        p.   cm.
    Includes bibliographical references and index.
    ISBN 0 471 94175 1   : — ISBN 0 471 94298 7 (pbk.)
    1. Cosmology.   I. Title.
QB981.R653   1994
523.1—dc20   93–32956
                                                        CIP

British Library Cataloguing in Publication Data

A catalogue record for this book is available from the British Library

ISBN 0 471 94175 1 (cloth)
ISBN 0 471 94298 7 (paper)

Typeset in 10/12pt Times from author's disks by Text Processing Department,
John Wiley & Sons Ltd, Chichester
Printed and bound in Great Britain by Bookcraft (Bath) Ltd, Midsomer-Norton, Avon

To Jacqueline

Contents

Preface

A few decades ago, astronomy and particle physics started to merge in the common field of cosmology. The general public had always been more interested in the visible objects of astronomy than in invisible atoms, and probably met cosmology first in Steven Weinberg's famous book *The First Three Minutes*. More recently Stephen Hawking's *A Brief History of Time* has caused an avalanche of interest in this subject.

Although there are now many popular monographs on cosmology, there are so far no introductory textbooks at university undergraduate level. Chapters on cosmology can be found in introductory books on relativity or astronomy, but they cover only part of the subject. One reason may be that cosmology is explicitly cross-disciplinary, and therefore it does not occupy a prominent position in either physics or astronomy curricula.

At the University of Helsinki I decided to try to take advantage of the great interest in cosmology among the younger students, offering them a one-semester course about one year before their specialization started. Hence I could not count on much familiarity with quantum mechanics, general relativity, particle physics, astrophysics or statistical mechanics. At this level, there are courses with the generic name of Structure of Matter dealing with Lorentz transformations and the basic concepts of quantum mechanics. My course aimed at the same level. Its main constraint was that it had to be taught as a one-semester course, so that it would be accepted in physics and astronomy curricula. The present book is based on that course, given three times to physics and astronomy students in Helsinki.

Of course there already exist good books on cosmology. The reader will in fact find many references to such books, which have been an invaluable source of information to me. The problem is only that they address a postgraduate audience that intends to specialize in cosmology research. My readers will have to turn to these books later when they have mastered all the professional skills of physics and mathematics.

In this book I am not attempting to teach basic physics to astronomers. They will need much more. I am trying to teach just enough physics to be able to explain the main ideas in cosmology without too much hand-waving. I have tried to avoid the other extreme, practised by some of my particle physics colleagues, of writing books on cosmology with the obvious intent of making particle physicists out of every theoretical astronomer.

I also do not attempt to teach basic astronomy to physicists. In contrast to astronomy scholars, I think the main ideas in cosmology do not require very detailed knowledge of astrophysics or observational techniques. Whole books have been written on distance measurements and the value of the Hubble parameter, which still remains imprecise to a factor of two. Physicists only need to know that quantities entering formulae are measurable — albeit incorporating factors h to some power — so that the laws can be discussed meaningfully. At undergraduate level, it is not even usual to give the errors on measured values.

In most chapters there are subjects demanding such a mastery of theoretical physics or astrophysics that the explanations have to be qualitative and the derivations meagre, for instance in general relativity, spontaneous symmetry breaking, inflation and galaxy formation. This is unavoidable because it just reflects the level of undergraduates. My intention is to go just a few steps further in these matters than do the popular monographs.

I am indebted in particular to two colleagues and friends who offered constructive criticism and made useful suggestions. The particle physicist Professor Kari Enqvist of NORDITA, Copenhagen, my former student, has gone to the trouble of reading the whole manuscript. The space astronomer Professor Stuart Bowyer of the University of California, Berkeley, has passed several early mornings of jet lag in Lapland going through the astronomy-related sections. Anyway, he could not go out skiing then because it was either a snow storm or $-30°$C! Finally, the publisher provided me with a very knowledgeable and thorough referee, an astrophysicist no doubt, whose criticism of the chapter on galaxy formation was very valuable to me. For all remaining mistakes I take full responsibility. They may well have been introduced by me afterwards.

Thanks are also due to friends among the local experts: particle physicist Professor Masud Chaichian and astronomer Professor Kalevi Mattila have helped me with details and have answered my questions on several occasions. I am also indebted to several people who helped me to assemble the pictorial material: Drs Subir Sarkar in Oxford, Rocky Kolb in the Fermilab, Carlos Frenk in Durham, Werner Kienzle at CERN, my assistant Jari Pennanen in Helsinki and members of the COBE team. All the new pictures were drawn by Merica Ahjolinna.

Finally, I must thank my wife Jacqueline for putting up with almost two years of near absence and full absentmindedness while writing this book.

M.R.

1 From Newton to Hubble

The history of ideas on the structure and origin of the Universe shows that mankind has always put itself at the centre of Creation. As astronomical evidence has accumulated, these anthropocentric convictions have had to be abandoned one by one. From the natural idea that the solid Earth is at rest and the celestial objects all rotate around us, we have come to understand that we inhabit an average-sized planet orbiting an average-sized sun, that the solar system is in the periphery of a rotating galaxy of average size, flying at hundreds of kilometres per second towards an unknown goal in an immense Universe, containing billions of similar galaxies.

Although the history of cosmology is long and fascinating, we shall not trace it in detail, nor any further back than to Newton, accounting in Section 1.1 only for those ideas which have fertilized modern cosmology directly, or which happened to be right although they failed to earn timely recognition.

Having a rigid Earth to stand on is a very valuable asset. How can we describe motion except in relation to a fixed point. Important understanding has come from the study of inertial systems, in uniform motion with respect to one another. From the work of Einstein on inertial systems, the theory of special relativity was born. In Section 1.2 we shall cover some aspects of this theory.

A classic problem is why the night sky is dark and not blazing like the disc of the Sun, as simple theory in the past would have it. In Section 1.3 we shall discuss this so called Olbers' paradox, and the modern understanding of it.

The beginning of modern cosmology may be fixed at the publication in 1929 of Hubble's law, which was based on observations of the red shift of spectral lines from remote galaxies. This was subsequently interpreted as evidence for the expansion of the Universe, thus ruling out a static Universe and thereby setting the primary requirement on theory. This will be explained in Section 1.4, in which we shall also discuss the implications of Hubble's law for our knowledge of the age and the present size of the Universe.

1.1 Pre-relativistic Cosmology

At the time of Isaac Newton (1642–1727) the heliocentric Universe of Nicolaus Copernicus (1473–1543), Galileo Galilei (1564–1642) and Johannes Kepler (1571–1630) had been accepted. Mankind was dethroned to live on an average size planet orbiting around an average size sun. The stars were understood to be suns like ours with fixed positions in a static Universe. The Milky Way had been resolved into an accumulation of faint stars with the telescope of Galileo. The *anthropocentric view* still persisted, however, in locating the solar system at the centre of this Universe.

The first theory of gravitation appeared when Newton published his *Philosophiae Naturalis Principia Mathematica* in 1687. With this theory he could explain the empirical laws of Kepler, that the planets moved in elliptical orbits with the Sun at one of the focal points. An early success of this theory came when Edmund Halley (1656–1742) successfully predicted that the comet sighted in 1456, 1531, and 1607 would return in 1758. In our time Newton's theory of gravitation still suffices to describe most of planetary and satellite mechanics, and it constitutes the non-relativistic limit of Einstein's relativistic theory of gravitation.

Newton considered the stars to be suns evenly distributed throughout infinite space in spite of the obvious concentration of stars in the Milky Way. A distribution is called *homogeneous* if it is uniformly distributed, and it is called *isotropic* if it has the same properties in all spatial directions. Thus in a homogeneous and isotropic space the distribution of matter would look the same to observers located anywhere; no point would be preferential. Clearly, matter introduces lumpiness which grossly violates homogeneity on the scale of stars, but on some larger scale isotropy and homogeneity may still be a good approximation. Going one step further, one may postulate what is called the *cosmological principle*, or sometimes the *Copernican principle*:

> The Universe is homogeneous and isotropic in three-dimensional space, has always been so, and will always remain so.

Whether this principle is true on any scale is not clear. On the galactic scale matter is lumpy, and galaxies tend to cluster and form what appears as narrow filaments separated by huge voids, as can be seen from Fig. 1 (Plate 2, frontispiece) [1]. One can quantify the homogeneity of matter in the Universe by the ratio $\delta N/N$ where N is the count of galaxies inside a randomly located sphere, and δN is the average (r.m.s.) fluctuations in N. The small scale lumpiness gives way to homogeneity only when the sphere becomes sufficiently large, thus $\delta N/N \approx 0.5$ for a radius of about 100 million light years. This is about the size of the supergalaxy in the constellation of Virgo to which our Galaxy and our local galaxy group belong. However, as new clustering is discovered on still larger scales, this number may need revision. In any case, the cosmological principle has been at the foundation of all cosmologies because it offers a useful simplification and approximation of the real world.

Based on his theory of gravitation, Newton formulated a cosmology in 1691. Since all massive bodies attract each other, a finite system of stars distributed over a finite region of space should collapse under their mutual attraction. But this was not observed, in fact the stars were known to have had fixed positions since antiquity, and Newton sought a reason for this stability. He concluded, erroneously, that the self-gravitation within a finite system of stars would be compensated for by the attraction of a sufficient number of stars outside the system, distributed evenly throughout infinite space. However, the total number of stars could not be infinite because then their attraction would also be infinite, making the static Universe unstable. It was understood only much later that the addition of external layers of stars would have no influence on the dynamics of the interior. The right conclusion is that the Universe cannot be static, an idea which would have been too revolutionary at the time.

Newton's contemporary and competitor Gottfried Wilhelm von Leibnitz (1646–1716) also regarded the Universe to be spanned by an abstract infinite space, but in contrast to Newton he also maintained that the stars must be infinite in number and distributed all over space, otherwise the Universe would be bounded and have a centre, contrary to contemporary philosophy. Finiteness was considered equivalent to boundedness, and infinity to unboundedness.

The first description of the Milky Way as a rotating galaxy can be traced to Thomas Wright (1711–1786), who wrote *An Original Theory or New Hypothesis of the Universe* in 1750, suggesting that the stars are 'all moving the same way and not much deviating from the same plane, as the planets in their heliocentric motion do round the solar body'.

The galactic picture of Wright had a direct impact on Immanuel Kant (1724–1804). In 1755 Kant went a step further, suggesting that the diffuse nebulae which Galileo had already observed could be distant galaxies rather than nearby clouds of incandescent gas. This implied that the Universe could be homogeneous on the scale of galactic distances in support of the cosmological principle.

Kant also pondered over the reason for transversal velocities such as the movement of the Moon. If the Milky Way was the outcome of a gaseous nebula contracting under Newton's law of gravitation, why was not all movement directed towards a common centre? Perhaps there also existed repulsive forces of gravitation which would scatter bodies onto other than radial trajectories, and perhaps such forces at large distances would compensate for the infinite attraction of an infinite number of stars? Note that the idea of a contracting gaseous nebula constituted the first example of a non-static system of stars, but at galactic scale with the Universe still static.

Kant also thought that he had settled the argument between Newton and Leibnitz about the finiteness or infiniteness of the system of stars. He claimed that either type of system embedded in an infinite space could not be stable and homogeneous, and thus the question of infinity was irrelevant. Similar thoughts can be traced to the scholar Yang Shen in China at about the same time, then unknown to Western civilisation [2].

The infinity argument was, however, not properly understood until Bernhard Riemann (1826–1866) pointed out that the world could be *finite*, yet *unbounded* provided the geometry of the space had a positive curvature, however small. On the basis of Riemann's geometry Albert Einstein (1879–1955) subsequently established the connection between the geometry of space and the distribution of matter.

Kant's repulsive force would have produced trajectories in random directions, but all the planets and satellites in the solar system exhibit transversal motion in one and the same direction. This was noticed by Pierre Simon de Laplace (1749–1827) who refuted Kant's hypothesis by a simple probabilistic argument in 1825: the observed movements were just too improbable if they were due to random scattering by a repulsive force. Laplace also showed that the large transversal velocities and their direction had their origin in the rotation of the primordial gaseous nebula and the law of conservation of angular momentum. Thus no nebula could contract to a point, and the Moon would not be expected to fall down upon us.

This leads to the question of the origin of time: what was the first cause of the rotation of the nebula and when did it all start? This is the question modern cosmology attempts to answer by tracing the evolution of the Universe backwards in time.

Note that although no repulsive force is needed to explain the transversal motion of the planets and their moons, this idea keeps coming back into modern cosmology in the form of a cosmological constant needed for other purposes.

The implications of Newton's gravity were quite well understood by John Michell (1724–1793) who pointed out in 1783 that a sufficiently massive and compact star would have such a strong gravitational field that nothing could escape from its surface [3]. This was the first mention of a type of star much later to be called a *black hole*. A few years after Michell, Laplace made the same observation.

Newton should also be credited with the invention of the reflecting telescope, which was only built one century later by William Herschel (1738–1822). With this instrument observational astronomy took a big leap forward: Herschel and his son John could map the nearby stars well enough in 1785 to conclude correctly that the Milky Way was a disc-shaped star system. They also concluded erroneously that the solar system was at its centre, but many more observations were needed before it was corrected. Herschel made many important discoveries, among them the planet Uranus, and some 700 binary stars whose movements confirmed the validity of Newton's theory of gravitation outside the solar system. He also observed some 250 diffuse nebulae which he first believed were distant galaxies, but which he and many other astronomers later considered to be nearby incandescent gaseous clouds belonging to our Galaxy. The main problem was then to explain why they avoided the directions of the galactic disc, since they were evenly distributed in all other directions.

The view of Kant that the nebulae were distant galaxies was defended also by Johann Heinrich Lambert (1728–1777). He was led to the conclusion that the solar system along with the other stars in our Galaxy orbited around the galactic centre, thus departing from the heliocentric view. The correct reason for the absence of nebulae in the galactic plane was only given by Richard Anthony Proctor (1837–1888), who proposed the presence of interstellar dust. The arguments for or against distant galaxies nevertheless raged throughout the nineteenth century until Edwin P. Hubble (1889–1953) in 1925 gave indisputable proof of the former view by determining the distance to several galaxies, among them the celebrated M31 galaxy in the Andromeda. Although this distance was off by a factor of two, the conclusion was qualitatively correct.

In spite of the work of Kant and Lambert the heliocentric picture of the Galaxy—or almost heliocentric since the Sun was located quite close to Herschel's galactic centre—remained well into our century. A decisive change came with the observations in 1915–1919 by Harlow Shapley (1895–1972) of the distribution of globular clusters. He found that perpendicular to the galactic plane they were uniformly distributed, but along the plane these clusters had a distribution which peaked in the direction of the Sagittarius. This defined the centre of the Galaxy to be quite far from the solar system: we are at a distance of about two thirds of the galactic radius. Thus the anthropocentric world picture received its second blow—and not the last one—if we count Copernicus' heliocentric picture as the first one. Note that Shapley still believed our Galaxy to be at the centre of the astronomical Universe.

Astronomical distance measurements were becoming more accurate at this time, building on a ladder of stars of known properties, so-called 'standard candles'. The *luminosity L* and *brightness* Φ of a star can be related to distance r according to

$$\Phi = \frac{L}{4\pi r^2} \; . \tag{1.1}$$

Luminosity is defined as the amount of luminous energy radiated per unit time, and brightness or *flux* as luminosity per unit surface.

Knowledge of astrophysical processes permits one to infer luminosity from observations of surface temperature and colour. Having established the distances to several stars in a cluster and their spectra, one can study the statistical frequency of stars of given luminosity and thereby establish cluster properties. The next rung of the ladder is to compare this cluster with similar clusters and galaxies further away.

In 1924 Hubble had measured the distances to nine spiral galaxies, and he found that they were extremely far away. The nearest one, M31 in the *Andromeda*, is now known to be at a distance of 20 galactic diameters (Hubble's value was about 8) and the farther ones at hundreds of galactic diameters. These observations established that the spiral nebulae are, as Kant had conjectured, stellar systems comparable in mass and size with the Milky Way, and that their spatial distribution confirmed the expectations of the cosmological principle on the scale of galactic distances.

To give the reader an idea of where in the Universe we are, what is nearby and what is far away some cosmical distances are listed in Table 1. On a cosmological scale we are not really interested in objects smaller than a galaxy! We generally measure cosmic distances in *parsec* units pc (prefixed k for 10^3 and M for 10^6). A parsec is the distance at which one second of arc is subtended by a length equalling the mean distance between

Table 1   Cosmic distances and dimensions

Distance to the Sun	$8'$ $15''$ (light minutes)
Distance to the nearest star (α Centauri)	1.3 pc
Diameters of globular clusters	5–30 pc
Thickness of our Galaxy, the 'Milky Way'	0.3 kpc
Distance to our Galactic centre	8 kpc
Radius of our Galaxy, the 'Milky Way'	12.5 kpc
Distance to the nearest galaxy (Large Magellanic Cloud)	55 kpc
Distance to the Andromeda nebula (M31)	770 kpc
Size of galaxy groups	1–5 Mpc
Thickness of filament clusters	5/h Mpc
Distance to the Local Supercluster centre (in Virgo)	20 Mpc/h
Distance to the 'Great Attractor'	44 Mpc/h
Size of superclusters	$\gtrsim$50/h Mpc
Size of large voids	60/h Mpc
Length of filament clusters	100/h Mpc
Size of the 'Great Wall'	$>60\times 170h^{-2}$ Mpc^2
Hubble radius	3000/h Mpc

Table 2   Cosmological and astrophysical constants

Unit	Symbol	Value
Speed of light	c	299 792 458 m/s
Light year	ly	0.3066 pc = 0.946×10^{16} m
Parsec	pc	3.261 ly = $3.085\ 677 \times 10^{16}$ m
Solar luminosity	$L_\odot$	3.826×10^{26} J/s
Solar mass	$M_\odot$	1.989×10^{30} kg
Solar equatorial radius	$R_\odot$	6.960×10^{8} m
Hubble parameter	H_0	100 h km/s Mpc =h/9.8 Gyr
	h	0.5 – 0.85
Newtonian constant	G	$6.672\ 59 \times 10^{-11}$ m^3/kg s^2
Planck constant	$\hbar$	$6.582\ 122 \times 10^{-22}$ MeV s
Planck mass	$M_P = \sqrt{\hbar c/G}$	1.221×10^{19} GeV/c^2
Planck time	$t_P = \sqrt{\hbar G/c^5}$	5.31×10^{-44} s
Boltzmann constant	k	8.617×10^{-5} eV/K
Stefan-Boltzmann constant	$a = \pi^2 k^4/15\,\hbar^3 c^3$	7.56×10^{-16} J/m^{3}K^4
Critical density of Universe	$\rho_C = 3H_0^2/8\pi\,G$	$2.8 \times 10^{11}\ h^2\ M_\odot$/Mpc^3 = $10.6\ h^2$ GeV/m^3

the Sun and the Earth. The value of the constant h appearing in Table 1 and the parsec unit are given in Table 2 where the values of some useful cosmological and astrophysical constants are listed.

In 1926–1927 Bertil Lindblad (1895–1965) and Jan Hendrik Oort (1900–1992) verified Laplace's hypothesis that the Galaxy indeed rotated, and they determined the period to be 10^8 years and the mass to be about $10^{11} M_\odot$. The conclusive demonstration that the Milky Way is an average size galaxy, in no way exceptional or central, was given only in 1952 by Walter Baade. This we may count as the third breakdown of the anthropocentric world picture.

Let me finally mention in this context that Einstein firmly believed in a static Universe until he met Hubble in 1929 and was overwhelmed by the evidence for what was to be called Hubble's law.

The later history of cosmology until our days has been excellently summarized by Peebles [4].

1.2. Inertial Frames

Newton's first law—the law of inertia—states that a system on which no forces act is either at rest or in uniform motion. Such systems are called *inertial frames*. Accelerated or rotating frames are not inertial frames. Newton considered that 'at rest' and 'in motion' implicitly referred to an *absolute space* which was unobservable but which had a real existence independently of mankind.

In 1883 Ernst Mach (1838–1916) published a historical and critical analysis of mechanics in which he rejected Newton's concept of an absolute space, precisely because it was unobservable. Mach demanded that the laws of physics should be based only on concepts which could be related to observations. Since motion still had to be referred to some frame at rest, he proposed to replace absolute space by an idealized rigid frame of fixed stars. Thus 'uniform motion' was to be understood as motion relative to the whole Universe.

Although Mach clearly realized that all motion is relative, it remained to Einstein to take the full step of studying the laws of physics as seen by observers in inertial frames in relative motion with respect to each other. Before going to Einstein's theory of special relativity, let us digress to a non-relativistic argument about the static or non-static nature of the Universe which follows directly from the cosmological principle.

Consider an observer A in an inertial frame who measures the density of stars and their velocities in the space around him. Because of the homogeneity and isotropy of space, he would see the same mean density of stars (at one time t) in two different directions $\mathbf{r}$ and $\mathbf{r}'$,

$$\rho_A(\mathbf{r},t) = \rho_A(\mathbf{r}',t) .$$

Another observer B in another inertial frame, see Fig. 2, looking in the direction $\mathbf{r}$ from his location would also see the same mean density of stars,

$$\rho_B(\mathbf{r}',t) = \rho_A(\mathbf{r},t) .$$

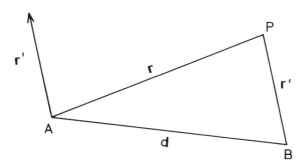

Fig. 2   Two observers at A and B making observations in the directions $\mathbf{r},\mathbf{r}'$

The velocity distributions of stars would also look the same to both observers, in fact in all directions, for instance in the $\mathbf{r}'$ direction,

$$\mathbf{v}_B(\mathbf{r}',t) = \mathbf{v}_A(\mathbf{r}',t) .$$

Suppose that the B frame has the relative velocity $\mathbf{v}_A(\mathbf{r}'',t)$ as seen from the A frame along the radius vector $\mathbf{r}'' = \mathbf{r} - \mathbf{r}'$. If all velocities are non-relativistic, i.e. small compared to the speed of light, we can write

$$\mathbf{v}_A(\mathbf{r}',t) = \mathbf{v}_A(\mathbf{r} - \mathbf{r}'',t) = \mathbf{v}_A(\mathbf{r},t) - \mathbf{v}_A(\mathbf{r}'',t) \ .$$

This equation is true only if $\mathbf{v}_A\ (\mathbf{r},t)$ has a specific form: it must be proportional to $\mathbf{r}$

$$\mathbf{v}_A(\mathbf{r},t) = f(t)\mathbf{r} \ , \qquad\qquad (1.2)$$

where $f(t)$ is an arbitrary function.

If $f(t) > 0$ the Universe would be seen by both observers to expand, each star having a radial velocity proportional to its distance $\mathbf{r}$. If $f(t) < 0$ the Universe would be seen to contract with velocities in the reversed direction. Thus we have seen that expansion and contraction are natural consequences of the cosmological principle.

An important lesson may be learned from studying the limited gravitational system consisting of the Earth and rockets launched into space. Suppose the rockets have initial velocities low enough to make them fall back onto Earth. The rocket–Earth gravitational system is then *closed* and contracting, corresponding to $f(t) < 0$.

When the kinetic energy is large enough to balance gravity the rockets satellize, they start to circulate Earth in stable Keplerian orbits at various altitudes. This requires the launch velocity to exceed 8 km/s, sometimes called the *first cosmic velocity*. This corresponds to the static solution $f(t) = 0$ for the rocket–Earth gravitational system.

If the launch velocities are increased beyond about 11 km/s, the potential energy of Earth's gravitational field no longer suffices to keep the rockets bound in Keplerian orbits. Beyond this speed, sometimes called the *second cosmic velocity*, the rockets escape for good. This is an expanding or *open* gravitational system, corresponding to $f(t) > 0$.

The intermediate case of satellites in Keplerian orbits is different if we consider the Universe as a whole. According to the cosmological principle, no point is preferred, and therefore there exists no centre around which bodies can gravitate in steady-state orbits. Thus the Universe is either expanding or contracting, the static solution being unstable and therefore unlikely.

Let us next study how signals are exchanged between inertial frames. I assume that the reader is familiar with Lorentz transformations and the basics of *special relativity*. Recall that the scope of the theory is to redefine the laws of physics in terms of *invariants* having the same value in all inertial frames.

A cornerstone in special relativity is the postulate that light travels by constant speed c in all frames. No material particle can be accelerated beyond c, no physical effect can be propagated faster than c and no signal can be transmitted faster than c. It is an experimental fact that no particle has been found travelling at superluminal speed, but a name for such particles has been invented, *tachyons*. Special relativity does not forbid tachyons, but if they exist they cannot be retarded to speeds below c. In this sense the speed of light constitutes a two-way barrier, an upper limit for ordinary matter and a lower limit for tachyons.

The infinitesimal time interval dt and the Pythagorean spatial distance element in three-dimensional space,

$$\mathrm{d}l = \sqrt{\mathrm{d}x^2 + \mathrm{d}y^2 + \mathrm{d}z^2} \ , \qquad\qquad (1.3)$$

are not invariants under Lorentz transformations, their value being different to observers in different inertial frames. Non-relativistic physics uses these quantities as completely adequate approximations, but in relativistic frame-independent physics we must find invariants to replace them.

This is possible if we replace ordinary three-dimensional space by the four-dimensional spacetime of Hermann Minkowski (1864–1909), defined by the spatial coordinates *x, y, z* and the temporal distance *ct*. The spatial distance d*l* between two space points in three-dimensional space is then generalized to a four-dimensional *spacetime distance* d*s* between two *events*, defined by the spatial coordinates as well as by the time coordinates. Scalar products in this four-dimensional space are invariants under *Lorentz transformations*.

It follows from the constancy of the speed of light that the scalar product of d*s* by itself is an invariant,

$$ds^2 = c^2 d\tau^2 = c^2 dt^2 - dx^2 - dy^2 - dz^2 = c^2 d\tau^2 - |dr|^2. \qquad (1.4)$$

The quantity dτ is called the *proper time* . Note that scalar multiplication in this manifold is defined in such a way that the products of the spatial components obtain negative signs.

The trajectory of a body moving in spacetime is called its *world line*. A body at a fixed location in space follows a world line parallel to the time axis and, of course, in the direction of increasing time. A body moving in space follows a world line making a slope with respect to the time axis. Since the speed of a body or a signal travelling from one event to another cannot exceed the speed of light, there is a maximum slope to such world lines. All world lines arriving where we are, here and now, obey this condition. Thus they form a cone in our past, and the envelope of the cone corresponds to signals travelling with the speed of light. This is called the *light cone*.

Two separate events in spacetime can be *causally* connected provided their spatial separation **dl** and their temporal separation d*t* (in any frame) obey

$$|\mathbf{dl}/dt| \leq c \ .$$

Their world line is then inside the light cone. One usually pictures this four-dimensional cone in *ct,r* space or (*ct,x,y*) space, see Fig. 3. Thus if we locate the *present* event at the apex of the light cone at $t = t_0 = 0$, it can be influenced by world lines from all events inside the *past* light cone for which $ct < 0$, and it can influence all events inside the *future* light cone for which $ct > 0$. Events inside the light cone are said to have *timelike* separation from the present event. Events outside the light cone are said to have *spacelike* separation from the present event: they cannot be causally connected to it. Thus the light cone encloses the *presently observable universe* that consists of all world lines that can in principle be observed. From now on we usually mean the presently observable universe when we say simply 'the Universe'.

For light signals the equality sign above applies so that the proper time interval in Eq. (1.4) vanishes,

$$d\tau = 0 \ .$$

Events on the light cone are said to have *null* or *lightlike* separation.

Let us now find out how the two observers measure velocities. The light emitted by stars originates in atomic transitions with emission spectra containing sharp spectral

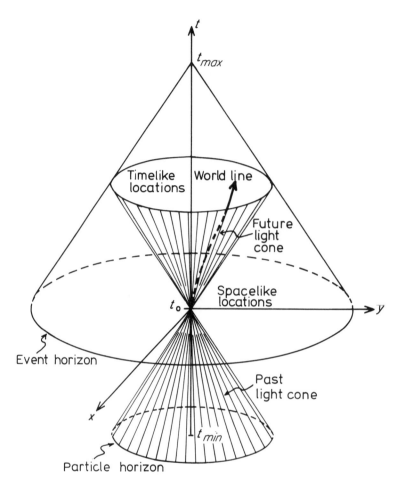

Fig. 3   Light cone in x,y,t-space. An event which is at the origin $x=y=0$ at the present time t_0 will follow some world line into the future, remaining always inside the future light cone. All points on the world line are at spacelike locations with respect to the spatial origin at t_0. World lines for light signals emitted from (received at) the origin at t_0 will propagate on the envelope of the future (past) light cone. No signals can be sent to or received from spacelike locations. The space in the past from which signals can be received at the origin at the present is restricted by the particle horizon at t_{min}, the earliest time under consideration. The event horizon restricts the space at the present time which can be in causal relation to the present spatial origin at some future time t_{max}

lines. Similarly, hot radiation traversing cooler matter in stellar atmospheres excites atoms at sharply defined wavelengths, producing characteristic dark absorption lines in the continuous regions of the emission spectrum. The observation of sharp, monochromatic absorption and emission lines constitute a measurement of the relative velocity of the source, mainly because radiation emitted with a wavelength λ_{rest} from a source moving away from us with velocity v will have its wavelength Doppler-shifted to λ_{obs} when observed on Earth. For a receding source this shift is in the red direction, $\lambda_{obs} > \lambda_{rest}$, and it is therefore called a *redshift*, denoted

$$z = \frac{\lambda_{obs} - \lambda_{rest}}{\lambda_{rest}} \;. \tag{1.5}$$

For an approaching source z is negative, corresponding to a *blueshift*.

If the radial recession velocity v of the source is very high—a non-negligible fraction of the speed of light—we should use special relativity to relate the wavelength shift z to v. Now since $d\tau^2$ is an invariant, it has the same value in the frame of a moving source of radiation (primed coordinates),

$$d\tau'^2 = d\tau^2 \;. \tag{1.6}$$

An observer at rest using a radiation clock records consecutive ticks separated by a spacetime interval $d\tau = dt = 1/v_{rest}$, where the frequency of the radiation, v_{rest}, is related to the wavelength by $v_{rest} = c/\lambda_{rest}$. An observer on Earth seeing such a clock receding with velocity v in the x' direction will observe that the ticks are separated by a time interval dt' and also by a space interval $dx' = vdt'$. From Eq. (1.6)

$$d\tau = d\tau' = \sqrt{dt'^2 - dx'^2/c^2} = \sqrt{1 - (v/c)^2} \; dt'.$$

In other words, the two inertial coordinate systems are related by a Lorentz transformation

$$dt' = \frac{dt}{\sqrt{1 - (v/c)^2}}. \tag{1.7}$$

Obviously the time interval dt' is always longer than the interval dt, but only noticeably so when v approaches c.

To the relativistic dilatation we now add the Doppler effect, well known from nonrelativistic physics. In the time interval dt' the distance from the radiation source to the observer will have increased from cdt' to $(c + v)dt'$, hence the total time interval between two received ticks in the terrestrial laboratory will be

$$\Delta t = (1 + v/c)dt' = \frac{(1 + v/c)dt}{\sqrt{1 - (v/c)^2}} \;.$$

The ratio of wavelengths actually measured by the terrestrial observer is then

$$1 + z = \frac{\lambda_{obs}}{\lambda_{rest}} = \frac{v_{rest}}{v_{obs}} = \frac{\Delta t}{dt} = \frac{(1 + v/c)}{\sqrt{1 - (v/c)^2}} \;. \tag{1.8}$$

For non-relativistic recession velocities, $v \ll c$, this is obviously well approximated by

$$z \simeq v/c. \tag{1.9}$$

This is illustrated in Fig. 4 where the dependence of z on v/c (multiplied by 3 Gpc) is plotted for the non-relativistic case (1.9) and for the relativistic case (1.8).

For relativistic recession velocities, $v \gtrsim c$, the connotation 'Doppler effect' is somewhat misleading. It is better to note simply that the wavelength λ_{obs} has been increased by the same factor $1 + z$ as the Universe has expanded between emission and absorption.

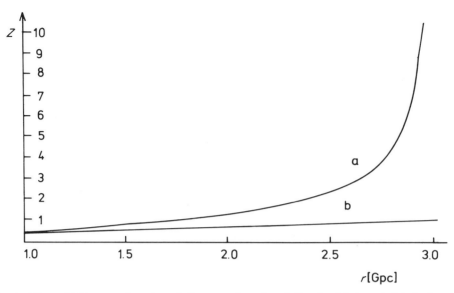

Fig. 4   The redshift z as a function of distance r in units of Gpc for (a) the relativistic formula (1.8) and (b) the non-relativistic approximation (1.9). The relation between r and v/c is given by Hubble's law (1.19) with $H_0^{-1} = 3$ Gpc

1.3 Olbers' Paradox

Let us turn to an early problem still discussed today, which is associated with the name of Wilhelm Olbers (1758–1840), although it seems to have been known already to Kepler in the 17th century, and a treatise on it was published by Jean-Philippe Loys de Chéseaux in 1744[5]. Why is the night sky dark if there are infinitely many stars evenly distributed in an infinite space? They should fill up the total field of visibility so that the night sky would be as bright as the Sun, and we would find ourselves in the middle of a heat bath of the temperature of the surface of the Sun. This question is obviously related to the number of shining stars, pondered already by Newton and Leibnitz.

Let us follow in some detail the argument published by Olbers in 1823. Suppose that the number of stars with average luminosity L is N and their average density in a volume V is $n = N/V$. If the average surface of a star is A, then its brightness is, by definition, $\Phi = L/A$. The Sun may be taken to be such an average star, mainly because we know it so well.

The number of stars in a spherical shell of radius r and thickness $\mathrm{d}r$ is then $4\pi r^2 n\,\mathrm{d}r$. Their total radiation as observed at the origin of a static universe of infinite extent is then found by integrating the spherical shells from 0 to infinity,

$$\int_0^\infty 4\pi r^2 n\Phi \,\mathrm{d}r = \int_0^\infty nL \,\mathrm{d}r = \infty. \tag{1.10}$$

On the other hand, a finite number of visible stars each taking up an angle A/r^2 could cover an infinite number of more distant stars, so it is not correct to integrate r to infinity. Let us integrate only up to such a distance R that the whole sky of angle 4π would be evenly tiled by the star discs. The condition for this is

$$\int_0^R 4\pi r^2 n \, \frac{A}{r^2} \, dr \; = \; 4\pi.$$

It then follows that the distance is $R = 1/An$. The integrated flux from these visible stars alone is then

$$\int_0^R nL \, dr \; = \; L/A, \tag{1.11}$$

or equal to the brightness of the Sun. But the night sky is indeed dark, so we are faced with a paradox.

Olbers' own explanation was that invisible interstellar dust absorbed the light. That would make the intensity of starlight decrease exponentially with distance. But one can show that the amount of dust needed would be so great that the Sun also would be obscured. Moreover, the radiation would heat the dust so that it would start to glow soon enough, thereby becoming visible in the infrared.

A large number of different solutions to this paradox have been proposed, some of the wrong ones persisting into our days. Let us here follow a valid line of reasoning due to Lord Kelvin (1824–1907), as retold and improved in a popular book by Harrison [5].

A star at distance r covers the fraction $A/4\pi r^2$ of the sky. Multiplying this with the number of stars in the shell, $4\pi r^2 n \, dr$, we obtain the fraction of the whole sky covered by stars viewed by an observer at the centre, $An \, dr$. Since n is the star count per volume element, An has the dimensions of number of stars per linear distance. The inverse of this,

$$\ell = 1/An \,, \tag{1.12}$$

is the mean radial distance between stars, or the *mean free path* of photons emitted from one star and being absorbed in collisions with another. We can also define a mean collision time,

$$\bar{\tau} = \ell/c \,. \tag{1.13}$$

The value of $\bar{\tau}$ can be roughly estimated from the properties of the Sun, with radius $R_\odot$ and density $\rho_\odot$. Let the present mean density of luminous matter in the Universe be ρ_0 and the distance to the farthest visible star r_*. Then the collision time inside this volume of size $4\pi r_*^3/3$ is

$$\bar{\tau} \simeq \bar{\tau}_\odot = \frac{1}{A_\odot nc} = \frac{1}{\pi R_\odot^2} \frac{4\pi r_*^3}{3Nc} = \frac{4\rho_\odot R_\odot}{3\rho_0 c} \,. \tag{1.14}$$

Taking the solar parameters from Table 2 (page 6) we obtain approximately 10^{23} years.

The probability that a photon does not collide, but arrives safely to be observed by us after a flight distance r can be derived from the assumption that the photon encounters obstacles randomly, that the collisions occur independently and at a constant rate ℓ^{-1} per unit distance. The probability $P(r)$ that the distance to the first collision is r is then given by the exponential distribution

$$P(r) = \ell^{-1}e^{-r/\ell} \ . \tag{1.15}$$

Thus flight distances much longer than ℓ are improbable.

Applying this to photons emitted in a spherical shell of thickness dr, and integrating the spherical shell from zero radius to r_*, the fraction of all emitted photons arriving at the centre of the sphere is

$$f(r_*) = \int_0^{r_*} \ell^{-1}e^{-r/\ell}dr = 1 - e^{-r_*/\ell} \ . \tag{1.16}$$

Obviously this fraction approaches 1 only in the limit of an infinite universe. In that case every point on the sky would be seen emitting photons, and the sky would indeed be as bright as the Sun at night. But since this is not the case, we must conclude that r_*/ℓ is small. Thus the reason why the whole field of vision is not filled with stars is that the volume of the presently observable Universe is too small to contain sufficiently many visible stars.

Lord Kelvin's original result follows in the limit of small r_*/ℓ, in which case

$$f(r_*) \approx r/\ell \ .$$

The exponential effect in Eq. (1.16) was neglected by Lord Kelvin.

We can also replace the mean free path in Eq. (1.16) by the collision time (1.13), and the distance r_* by the age of the Universe t_0 to obtain the fraction

$$f(r_*) = g(t_0) = 1 - e^{-t_0/\bar{\tau}} \ . \tag{1.17}$$

If $u_\odot$ is the average radiation density at the surface of the stars then the radiation density u_0 measured by us is correspondingly reduced by the fraction $g(t_0)$,

$$u_0 = u_\odot(1 - e^{-t_0/\bar{\tau}}) \ . \tag{1.18}$$

The condition to observe a luminous night sky is $u_0 \approx u_\odot$, or the Universe must have an age of the order of the collision time, $t_0 \approx 10^{23}$ years. However, this exceeds all estimates of the age of the Universe (some estimates will be given in the next section) by 13 orders of magnitude! Thus the existing stars have not had time to radiate long enough.

What Olbers and many after him did not take into account is that even if the age of the Universe were infinite, the stars do have a finite age and they burn their fuel at well understood rates.

If we replace 'stars' by 'galaxies' in the above argument, the problem changes quantitatively but not qualitatively. The intergalactic space is filled with radiation from the galaxies, but there is less of it than one would expect for an infinite Universe, at all wavelengths. There is still a problem to be solved, but it is not quite as paradoxical as in Olbers' case.

One explanation is the one we have already met: each star radiates only for a finite time, and each galaxy has existed only for a finite time, whether the age of the Universe is infinite or not. Thus when the time perspective grows, an increasing number of stars become visible because their light has had time to reach us, but at the same time stars which have burned their fuel disappear.

Another possible explanation evokes expansion and special relativity. If the Universe expands, starlight redshifts, so that each arriving photon carries less energy than when it was emitted. At the same time the volume of the Universe grows, and thus the energy density decreases. The observation of the low level of radiation in the intergalactic space has in fact been evoked as a proof of the expansion.

Since both explanations certainly contribute, it is necessary to carry out detailed quantitative calculations to establish which of them is more important. Most of the existing literature on the subject supports the relativistic effect, but Harrison has shown (and Wesson [6] has further emphasized) that this is false: the finite lifetime of the stars and galaxies is the dominating effect. The relativistic effect is quantitatively so unimportant that one cannot use it to prove that the Universe is either expanding or contracting.

1.4 Hubble's Law

In the 1920s Hubble measured the velocities of 18 spiral galaxies with a reasonably well-known distance. The expectation for a stationary universe was that galaxies would be found moving about randomly. However, some observations had already shown that most galaxies were redshifted, thus receding, although some of the nearby ones exhibited blueshift. For instance, the nearby Andromeda nebula is approaching us, as is testified by its blueshift. Hubble's fundamental discovery was that the velocities increased linearly with distance,

$$v = H_0 r. \tag{1.19}$$

This is called *Hubble's law* and H_0 the *Hubble parameter*. This law has been verified since then by the observations of some 30 000 galaxies out to redshifts of $z \approx 0.5$.

The value initially found by Hubble was $H_0 = 550$ km s^{-1} Mpc^{-1} (Mpc stands for megaparsec, see Table 2), but present values range from 50 to 85 in these units. Since nothing dictates that H_0 has been constant during all times rather than a function $H(t)$, one adds the subscript 0 to imply its value *now* at time t_0. It is common practice to write

$$H_0 = 100h \text{ km s}^{-1} \text{ Mpc}^{-1}, \tag{1.20}$$

and put the uncertainty into the constant h. This is what we have done in Table 2.

The message of Hubble's law is that the Universe is expanding. At a scale of tens or hundreds of Mpc all distances are increasing regardless of the position of our observation point. It is true that we observe that the galaxies are receding *from us* as if we were at the centre of the Universe. However, this is exactly what the expansion should look like to every observer if the cosmological principle is valid.

It is surprising that neither Newton nor later scientists, pondering about why the Universe avoided a gravitational collapse, came to realize the correct solution. An expanding universe would be slowed down by gravity, so the inevitable collapse would be postponed until later. Probably it was the notion of an infinite scale of time, inherent in a stationary model, which blocked the way to the right conclusion.

From Eqs. (1.19) and (1.20) one sees that the Hubble parameter has the dimension of inverse time. Thus a characteristic time scale for the expansion of the Universe is the *Hubble time*

$$\tau_H \equiv H_0^{-1} = 9.8h^{-1} \times 10^9 \text{yr.} \tag{1.21}$$

We shall see later that τ_H constitutes an upper limit to the age t_0 of the Universe in standard cosmological models. With h in the range 0.5–0.85, t_0 ranges in 11.5–19.6 Gyr.

There are many *cosmochronometers* yielding some information on t_0. The oldest reliably established ages of globular clusters in the spheroid of the Galaxy are 15 ± 3 Gyr, which gives $t_0 \approx 13 - 20$, allowing 1–2 Gyr from the Big Bang until the Galaxy forms these stars. The distribution in colour of He-burning stars in the bulge of the Galaxy seems to imply ages in the range 14–18 Gyr. White dwarf stars determine the age of the considerably younger galactic disc to be typically $\lesssim 10$ Gyr, but since we do not know how long after the Big Bang the disc was formed this is not a useful cosmochronometer.

Various nuclear processes can also be used to date the age of the Galaxy, t_G, for instance the 'Uranium clock'. Long-lived radioactive isotopes such as ^{232}Th, ^{235}U, ^{238}U, and ^{244}Pu have been formed by fast neutrons (from supernova explosions) captured in the envelopes of an early generation of stars. The proportions of these isotopes at the time of production are calculable with some degree of confidence. Since then, they have decayed with their different natural half-lives so that their abundances in the Galaxy today are changed, and well measurable. For instance, recent calculations of the original ratio $K = {}^{235}U/{}^{238}U$ give values between 1.16 and 1.34, whereas this ratio at the present time is $K_0 = 0.00723$.

To compute the age of the Galaxy by this method, we also need the decay constants λ of ^{238}U and ^{235}U which are related to their half-lives,

$$\lambda_{238} = \ln2/(4.46 \text{ Gyr}),$$
$$\lambda_{235} = \ln2/(0.7038 \text{ Gyr}).$$

The relation between isotope proportions, decay constants, and time t_G is

$$K = K_0 \, \exp[(\lambda_{238} - \lambda_{235})t_G] . \tag{1.22}$$

Inserting numerical values one finds

$$t_G \approx 6.2 \text{ Gyr.}$$

Detailed astrophysical considerations about star formation rates double this figure, but the uncertainty is easily of the order of 2 Gyr. Using all information (see e.g. references [7], [8]) one can only conclude that the age of the Universe is in the range

$$14 \text{ Gyr} \lesssim t_0 \lesssim 18 \text{ Gyr.} \tag{1.23}$$

Note that the very existence of radioactive nuclides testifies that the Universe could not have been infinitely old and static.

The Hubble parameter also determines the size scale of the presently observable Universe. In time τ_H, radiation travelling with the speed of light c has reached the *Hubble radius*

$$r_H \equiv \tau_H c = 3000h^{-1}\text{Mpc}. \tag{1.24}$$

Or, to put it differently, according to Hubble's law, objects at this distance recede with the speed of light which is an absolute limit in the theory of special relativity. Combining Eq. (1.19) with the non-relativistic expression (1.9) one obtains

$$z = H_0 \frac{r}{c} . \tag{1.25}$$

This is of course true only for $z \ll 1$ galaxies, since one can see from the relativistic expression (1.8) that z is infinite for objects at distance r_H receding with the speed of light (see Fig. 4) because the expression under the root sign vanishes. Therefore no information can reach us from farther away, all radiation is redshifted to infinite wavelengths and no particle emitted within the Universe can exceed this distance.

The size of the expanding Universe is best measured by a relative *cosmic scale factor* $S(t)$, the present value being denoted $S_0 \equiv S(t_0)$. We need not identify S_0 with r_H, rather it can be thought of as the distance from where we are to some redshifted object, the most distant one (in 1993) being a quasar at $z = 4.9$. Because of the expansion, a wavelength λ of light emitted at time $t < t_0$ has been scaled to λ_0 at time t_0,

$$\frac{\lambda_0}{S_0} = \frac{\lambda}{S(t)}.$$

Let us find an approximation for $S(t)$ by expanding it to first order in time differences,

$$S(t) \approx S_0 - \dot{S}(t_0)(t_0 - t) . \tag{1.26}$$

Using the notation $\dot{S}_0$ for $\dot{S}(t_0)$ and $r = c(t_0 - t)$ for the distance to the source, the redshift can be approximated by

$$z = \frac{\lambda_0}{\lambda} - 1 = \frac{S_0}{S} - 1 \approx \frac{\dot{S}_0}{S_0} \frac{r}{c} . \tag{1.27}$$

Identifying the expressions for z in Eqs. (1.27) and (1.25) we find the important relation

$$\frac{\dot{S}_0}{S_0} = H_0 , \tag{1.28}$$

which is true for $z \ll 1$.

The expansion can also be verified by measuring the surface brightness of 'standard sources' at varying redshifts $1+z$, the *Tolman test*. For sources showing redshift, regardless of whether its cause is expansion or something else, the energy of each received photon is reduced by a factor $1 + z$ (this is often called the *energy effect*).

If the Universe does indeed expand, the intensity of the photon signal at the detector is further reduced by another factor $1 + z$ due to the stretching of the path length, and by a factor $(1 + z)^2$ due to an optical aberration which makes the surface of the source appear increased. Such tests have been done and they do confirm the expansion.

Problems

1. The radius of the galaxy is 3×10^{20} m. How fast would a spaceship have to travel to cross it in 300 yr as measured on board? Express your result in terms of $\gamma = 1/\sqrt{1 - v^2/c^2}$. [9]

2. Quasistellar objects exhibit a Doppler shift of magnitude $z = (\lambda_{obs} - \lambda)/\lambda = 1.95$, where λ is the wavelength that would have been measured by an observer at rest relative to the emitter of the radiation. Use this information to calculate the velocity with which these objects move relative to us. How far away are they (in light years) if they obey Hubble's law? [9]

3. An observer sees a spaceship coming from the west at a speed of $0.6c$ and a spaceship coming from the east at a speed $0.8c$. The western spaceship sends a signal with a frequency of 10^4 Hz in its rest frame. What is the frequency of the signal as perceived by the observer? If the observer sends on the signal immediately upon reception, what is the frequency with which the eastern spaceship receives the signal? [9]

4. If the eastern spaceship in the previous problem were to interpret the signal as one that is Doppler shifted because of the relative velocity between the western and eastern spaceships, what would the eastern spaceship conclude about the relative velocity? Show that the relative velocity must be $(v_1 + v_2)/(1 + v_1 v_2/c^2)$, where v_1 and v_2 are the velocities as seen by an outside observer. [9]

5. A source flashes with a frequency of 10^{15} Hz. The signal is reflected by a mirror moving away from the source with speed 10 km/s. What is the frequency of the reflected radiation as observed at the source? [9]

6. Use Eq. (1.14) to estimate the mean free path ℓ of photons. What fraction of all photons emitted by stars up to the maximum known distance $z = 4.9$ arrive at Earth?

References

1. M. J. Geller and J. P. Huchra, *Science*, **246** (1989) 897.
2. Fang Li Zhi and Li Shu Xian, *Creation of the Universe*, World Scientific Publ. Co., Singapore, 1989.
3. S. W. Hawking, *A Brief History of Time*, Bantam Books, New York, 1988.
4. P. J. E. Peebles, *Principles of Physical Cosmology*. Princeton University Press, Princeton, New Jersey, 1993.
5. E. Harrison, *Darkness at Night*. Harvard University Press, Cambridge, Mass., 1987.
6. P. S. Wesson, *The Astrophysical Journal*, **367** (1991) 399.
7. F.-K. Thielemann in *A Unified View of the Macro- and Micro-cosmos*, First International School on Astroparticle Physics, Erice 1987. Eds. A. de Rújula *et al.*, World Scientific Publ. Co., Singapore, 1987, pp. 424.
8. A. Dressler in *Astronomy, Cosmology and Fundamental Physics*, Third ESO/CERN Symposium. Eds. M. Caffo *et al.*, Kluwer Academic Publ., Dordrecht, 1988, p. 23.
9. S. Gasiorowicz, *The Structure of Matter*. Addison-Wesley Publ. Co., Reading, Mass., 1979.

2 Laws of Gravitation

The laws of physics distinguish between several kinds of forces or interactions: gravitational, electroweak, strong. Formerly, there were the electromagnetic and weak interactions now united in the electroweak interaction, and even earlier the electric and the magnetic interactions were separate. It is a dream of physics to unite all these forces into a grand unified theory.

The weakest interaction is that of gravitation. Its weakness is well manifested by the fact that it takes bodies of astronomical size to make the gravitational interaction noticeable. Yet it is the most important one for understanding the Universe. The electromagnetic force is unimportant because the astronomical objects and the Universe as a whole are electrically neutral. The weak interactions have a range of 10^{-17} cm only, and the strong interactions about 10^{-13} cm, so they are important for particles on atomic scales, but not for astronomical bodies. The non-gravitational forces still do play a role during an early epoch in the history of the Universe when it consisted of particles in interaction. We shall return to this in Chapter 4.

The foundations of modern cosmology were laid during the second and third decade of this century: on the theoretical side by Einstein's theory of general relativity which represented a deep revision of current concepts, and on the observational side by Hubble's discovery of the cosmic expansion which ruled out a static Universe and set the primary requirement on theory.

In Section 2.1 we shall turn to Newton's theory of gravitation which is the earliest explanation of a gravitational force. We shall 'modernize' it by introducing Hubble's law into it. In fact, we shall see that this leads to a cosmology which already contains many features of current Big Bang cosmologies.

The main purpose of this chapter is to derive Einstein's law of gravitation using as few mathematical tools as possible (for far more detail, see e.g. refs [1] and [2]). For this we need some understanding of spacetime and of various possible metrics, in particular the Robertson–Walker metric in a four-dimensional manifold. This forms the subject of Section 2.2. The basic principle of covariance introduced in Section 2.3 requires a brief encounter with tensor analysis. The principle of equivalence is introduced in Section 2.4 and it is illustrated by examples of travels in elevators. In the final Section 2.5 we assemble all these tools and arrive at Einstein's law of gravitation.

2.1. Expansion in a Newtonian World

A system of massive bodies in an attractive Newtonian potential contracts rather than expands. The solar system has contracted to a stable, gravitationally bound configuration from some form of hot gaseous cloud, and the same mechanism is likely to be true for larger systems such as the Milky Way, and perhaps also for clusters of galaxies. On larger scales the Universe expands, but this does not contradict Newton's law of gravitation.

The key question in cosmology is whether the Universe as a whole is a gravitationally bound system in which the expansion will be halted one day. We shall next derive a condition for this from Newtonian mechanics.

Consider a galaxy of *gravitating mass* m_G located at a radius r from the center of a sphere of mean density ρ and mass $M = 4\pi r^3\rho/3$, see Fig. 5. The gravitational potential of the galaxy is

$$U = -\frac{GMm_G}{r} = -\frac{4\pi}{3}Gm_G\rho r^2 \, , \qquad (2.1)$$

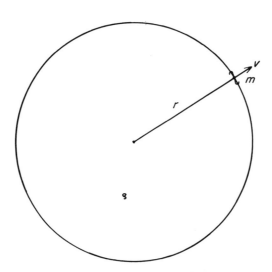

Fig. 5 A galaxy of mass m at radial distance r receding with velocity v from the center of a homogeneous mass distribution of density ρ

where G is the *Newtonian constant* expressing the strength of the gravitational interaction. Thus the galaxy falls towards the centre of gravitation, acquiring a radial acceleration

$$\ddot{r} = -\frac{GM}{r^2} = -\frac{4\pi}{3}G\rho r \, . \qquad (2.2)$$

This is *Newton's law of gravitation*, usually written in the form

$$F = -\frac{GMm_G}{r^2} \, , \qquad (2.3)$$

where F (in old-fashioned parlance) is the force exerted by the mass M on the mass m_G. The negative signs in the Eqs. (2.1–2.3) express the attractive nature of gravitation: bodies are forced to move in the direction of decreasing r.

In a universe expanding according to Hubble's law, the kinetic energy T of the galaxy receding with velocity v is

$$T = \tfrac{1}{2}mv^2 = \tfrac{1}{2}mH_0^2 r^2 \,, \tag{2.4}$$

where m is the *inert mass* of the galaxy. Although there is no theoretical reason for the inertial mass to equal the gravitational mass (we shall come back to this question later), careful tests have verified the equality to a precision better than one part in 10^{11}. Let us therefore set $m_G = m$. Thus the total energy is given by

$$E = T + U = \tfrac{1}{2}mH_0^2 r^2 - \frac{4\pi}{3}Gm\rho r^2 = mr^2 \left(\tfrac{1}{2}H_0^2 - \frac{4\pi}{3}G\rho \right) \,. \tag{2.5}$$

If the mass density ρ of the Universe is large enough, the expansion will halt. The condition for this to occur is $E = 0$, or from Eq. (2.5) this *critical density* is

$$\rho_c = \frac{3H_0^2}{8\pi G} \,. \tag{2.6}$$

A universe with density $\rho > \rho_c$ is called *closed*; with density $\rho < \rho_c$ it is called *open*.

Note that r and ρ are time-dependent: they scale with the the expansion. Denoting their present values r_0 and ρ_0, one has

$$\frac{r(t)}{r_0} = \frac{S(t)}{S_0} \;; \quad \frac{\rho(t)}{\rho_0} = \left(\frac{S_0}{S(t)} \right)^3 \,. \tag{2.7}$$

Since we do not know the radius of the Universe very well, it is more convenient to express all dynamical equations in terms of the scale factor S. The acceleration $\ddot{r}$ in Eq. (2.2) can then be replaced by the acceleration of the scale,

$$\ddot{S} = \frac{S_0}{r_0}\ddot{r} = -\frac{4\pi}{3}G\rho_0 S_0^3 \cdot \frac{1}{S^2} \equiv -\frac{a}{S^2} \,, \tag{2.8}$$

where a is a constant introduced for later use. Let us use the identity

$$\tfrac{1}{2}\frac{d}{dt}\dot{S}^2 = \dot{S}\ddot{S} = \frac{dS}{dt}\ddot{S}$$

to replace $\ddot{S}$ above by the expression

$$\ddot{S} = \tfrac{1}{2}\frac{d}{dS}\dot{S}^2 \,.$$

It then follows from Eq. (2.8) that

$$d\dot{S}^2 = -\frac{2a}{S^2}dS \,.$$

This can be integrated from the present time t_0 to an earlier time t,

$$\int_{t_0}^{t} d\dot{S}^2 = -2a \int_{S_0}^{S} \frac{dS}{S^2} ,$$

with the result

$$\dot{S}^2 - \dot{S}_0^2 = 2a(S^{-1} - S_0^{-1}) . \tag{2.9}$$

Let us now introduce the dimensionless *density parameter*

$$\Omega_0 = \frac{\rho_0}{\rho_c} = \frac{8\pi G \rho_0}{3H_0^2} , \tag{2.10}$$

where the subscripts 0 indicate present values, as usual. In terms of Ω_0 the constant a in Eq. (2.8) can be written

$$a = \frac{4\pi}{3} G S_0^3 \rho_0 = \tfrac{1}{2} H_0^2 S_0^3 \Omega_0 . \tag{2.11}$$

Substituting this expression for a and making use of the relation $\dot{S}_0 = S_0 H_0$ we find

$$\dot{S}^2 = S_0^2 H_0^2 + H_0^2 S_0^3 \Omega_0 \frac{1}{S_0} \left(\frac{S_0}{S} - 1 \right) = S_0^2 H_0^2 \left(1 - \Omega_0 + \Omega_0 \frac{S_0}{S} \right) . \tag{2.12}$$

During expansion $\dot{S}$ is positive; during contraction negative. In both cases the value of $\dot{S}^2$ is non-negative, so it must always be true that

$$\Omega_0 \frac{S_0}{S(t)} - \Omega_0 + 1 \geq 0 . \tag{2.13}$$

Depending on the value of Ω_0 the evolution of the Universe can take three courses:

(i) $\Omega_0 < 1$, the mass density is undercritical. As the cosmic scale factor $S(t)$ increases for times $t > t_0$ the term $\Omega_0 S_0 / S(t)$ decreases, but the expression (2.13) stays always positive. The rate of increase, $\dot{S}$ in Eq. (2.12), continues to increase since it scales with S,

$$\dot{S} = \dot{S}_0 \frac{S}{S_0} > \dot{S}_0 > 0 .$$

Thus this case corresponds to an open, ever expanding universe, as a consequence of the fact that it is expanding now. In Fig. 6 the expression in Eq. (2.13) is plotted against S.

(ii) $\Omega_0 = 1$, the mass density is critical. As the scale factor $S(t)$ increases for times $t > t_0$ the expression in Eq. (2.13) gradually approaches zero, and the expansion halts. However, this only occurs infinitely late, so this case also corresponds to an ever expanding universe. In Fig. 6 the cases (i) and (ii) differ by having different asymptotes.

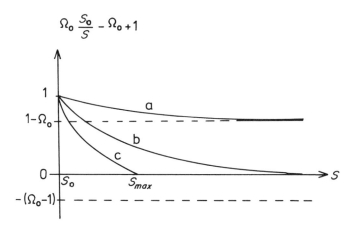

Fig. 6 Dependence of the expression (2.13) on the cosmic scale S for an (a) undercritical, (b) critical, (c) overcritical universe. Time starts at scale S_0 in this picture and increases with S, except for case (c) where the Universe arrives at its maximum size S_{max} at time t_{max}, whereupon it reverses its direction

(iii) $\Omega_0 > 1$, the mass density is overcritical and the Universe is closed. As the scale factor $S(t)$ increases, it reaches a value S_{max} at a time $t_{mid} > t_0$ when the expression in Eq. (2.13) vanishes, and the rate of increase, $\dot{S}$, also vanishes. But the condition (2.13) must stay true, and therefore the expansion must turn into contraction at time t_{mid}. For later times the Universe retraces the curve in Fig. 6, ultimately reaching size S_0 at time $2\,t_{mid}$.

The observational value of Ω_0 ranges from 0.1 to 0.2 when one uses galaxies as test particles to make dynamical measurements of matter. The problem then is that one is sensitive to the gravitational mass clustered on a scale smaller than that sampled by the galaxies. If there exists mass which is spread out more homogeneously than galaxy clusters, it does not contribute to the above value. Thus it may still be possible to have $\Omega_0 = 1$ or even larger. We shall return to this question frequently.

In the previous chapter we related the rate of expansion to a Hubble parameter H_0 assumed independent of time. H_0 actually appeared as a dynamical parameter in the lowest order Taylor's expansion of $S(t)$, Eq. (1.26). If we allow some mild time dependence to $H(t)$, that would correspond to introducing another dynamical parameter along with the next term in the Taylor's expansion. Thus adding the second-order term to Eq. (1.26), we have

$$S(t) \approx S_0 - \dot{S}_0(t_0 - t) + \tfrac{1}{2}\ddot{S}_0(t_0 - t)^2 \; . \tag{2.14}$$

Let us now introduce the *deceleration parameter* q_0 defined by

$$q_0 = -\frac{S_0\ddot{S}_0}{\dot{S}_0^2} = -\frac{\ddot{S}_0}{S_0 H_0^2} . \tag{2.15}$$

Then the lowest order expression for the redshift, Eq. (1.27), is replaced by

$$z = (S_0/S - 1) \approx [1 - H_0(t - t_0) - \tfrac{1}{2}q_0 H_0^2(t - t_0)^2]^{-1} - 1$$

$$\approx H_0(t - t_0) + (1 + \tfrac{1}{2}q_0)H_0^2(t - t_0)^2 = H_0 r/c + (1 + \tfrac{1}{2}q_0)(H_0 r/c)^2 \; .$$

The above expression can be inverted to obtain

$$H_0 \frac{r}{c} \approx z + \tfrac{1}{2}(1 - q_0)z^2. \tag{2.16}$$

The first term on the right gives Hubble's linear law (1.25), and thus the second term measures deviations from linearity to lowest order. The parameter value $q_0 = 1$ obviously corresponds to no deviation.

The linear law has been used to determine H_0 from galaxies within the local Supergalaxy. On the other hand, one also observes deceleration of the expansion in the local universe due to the lumpiness of matter. For instance, the Local Group clearly feels the overdensity of the Virgo cluster at a distance of about 20 Mpc. The motions of galaxies within gravitationally bound systems are called *peculiar velocities*. It has been argued that the peculiar velocities in the Local Supergalaxy cannot be understood without the pull of the neighbouring Hydra-Centaurus supercluster and perhaps a still larger overdensity in the supergalactic plane, nicknamed 'the Great Attractor'.

It should be clear from this that one needs to go to even greater distances, beyond the influences of local overdensities, to determine a value for q_0. Within the Local Supergalaxy it is safe to conclude that only the linear term in Hubble's law is necessary.

This is as far as we can go combining Newtonian mechanics with Hubble's law. We have seen that problems appear when the recession velocities exceed the speed of light, conflicting with special relativity. Another problem is that Newton's law of gravitation knows no delays: the gravitational potential is felt instantaneously over all distances. A third problem with Newtonian mechanics is that the Copernican world, assumed homogeneous and isotropic, extends up to a finite distance r_0, but outside that boundary there is nothing. Then the boundary region is characterized by violent inhomogeneity and anisotropy which is not taken into account. To cope with these problems we must begin to construct a fully relativistic cosmology.

2.2 The Metric of Spacetime

In Newton's time the laws of physics were considered to operate in a *flat Euclidean space*, in which spatial distance could be measured on an infinite and immovable three-dimensional grid, and time was a parameter marked out on a linear scale running from infinitely past to infinite future. But Newton could not answer the question of how to identify which inertial frame was at rest relative to this absolute space. In his days the solar frame could have been chosen, but today we know that the solar system orbits the Galactic centre, the Galaxy is in motion relative to the local galaxy group which in turn is in motion relative to the Hydra-Centaurus cluster and the whole Universe is expanding.

Riemann then realized that Euclidean geometry was just a particular choice suited to flat space, but not necessarily correct in the space we inhabit. And Mach realized that one had to abandon the concept of absolute space altogether. Einstein finally drew the conclusion and replaced the flat Euclidean three-dimensional space with curved Minkowskian four-dimensional space in which physical quantities are described by invariants.

Let us consider how to describe distance in three-space. The path followed by a free body obeying Newton's first law of motion can suitably be described by expressing its spatial coordinates as functions of time: $x(t)$, $y(t)$, $z(t)$. Time is then treated as an absolute parameter and not as a coordinate. This path represents the shortest distance between any two points along it, and it is called a *geodesic* of the space. As is well known, in Euclidean space the geodesics are straight lines. Note that the definition of a geodesic does not involve any particular coordinate system.

The *metric equation* for the infinitesimal squared distance element dl^2 takes the Euclidean form (1.3) in Cartesian coordinates, but the same flat space could equally well be mapped by, for example, spherical or cylindrical coordinates. For instance, choosing spherical coordinates R,θ,ϕ as in Fig. 7,

$$x = R \sin\theta \cos\phi , \quad y = R \sin\theta \sin\phi , \quad z = R \cos\theta , \tag{2.17}$$

dl^2 takes the form

$$dl^2 = dR^2 + R^2 d\theta^2 + R^2\sin^2\theta \, d\phi^2 . \tag{2.18}$$

Geodesics in this space obey Newton's first law of motion which may be written

$$\ddot{\mathbf{R}} = 0 . \tag{2.19}$$

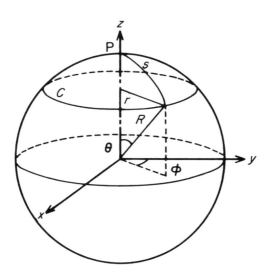

Fig. 7 A two-sphere on which points are specified by coordinates (θ,ϕ)

In special relativity, symmetry between spatial coordinates and time is achieved, as is evident from the *Minkowski metric* (1.4) describing a flat spacetime in four Cartesian coordinates. The path of a body, or its world line, is then described by the four coordinate functions $x(\tau)$, $y(\tau)$, $z(\tau)$, $t(\tau)$, where the proper time τ is a new absolute parameter, an invariant under Lorentz-transformations. A geodesic in the Minkowski spacetime is also a straight line, given by the equations

$$\frac{d^2 t}{d\tau^2} = 0, \quad \frac{d^2 \mathbf{R}}{d\tau^2} = 0 . \tag{2.20}$$

In the spherical coordinates (2.17) the Minkowski metric (1.4) takes the form

$$ds^2 = c^2 dt^2 - dl^2 = c^2 dt^2 - dR^2 - R^2 d\theta^2 - R^2 \sin^2\theta \, d\phi^2 . \tag{2.21}$$

An example of a curved space is the 2-dimensional surface of a sphere with radius S obeying the equation

$$x^2 + y^2 + z^2 = S^2. \tag{2.22}$$

This surface is called a two-sphere.

Combining Eqs. (1.3) and (2.22) we see that one coordinate is really superfluous, for instance z, so that the spatial metric (1.3) can be written

$$dl^2 = dx^2 + dy^2 + \frac{(xdx + ydy)^2}{S^2 - x^2 - y^2} . \tag{2.23}$$

This metric describes spatial distances on a two-dimensional surface embedded in three-space, but the third dimension is not needed to measure a distance on the surface. Note that S is not a third coordinate, but a constant everywhere on the surface.

Thus measurements of distances depend on the geometric properties of space, as has been known to navigators ever since Earth was understood to be spherical. The geodesics on a sphere are great circles, and the metric is

$$dl^2 = R^2 d\theta^2 + R^2 \sin^2\theta \, d\phi^2 . \tag{2.24}$$

Near the poles where $\theta = 0°$ or $\theta = 180°$ the local distance would depend very little on changes in longitude ϕ. No point on this surface is preferred, so it can correspond to a Copernican homogeneous and isotropic two-dimensional universe which is unbounded, yet finite.

Let us write Eq. (2.24) in the matrix form

$$dl^2 = (d\theta \ \ d\phi) \ \mathbf{g} \begin{pmatrix} d\theta \\ d\phi \end{pmatrix} , \tag{2.25}$$

where $\mathbf{g}$ is the metric matrix

$$\mathbf{g} = \begin{pmatrix} R^2 & 0 \\ 0 & R^2 \sin^2\theta \end{pmatrix} . \tag{2.26}$$

The 'volume' or area A of the two-sphere in Fig. 7 can then be written

$$A = \int_0^{2\pi} d\phi \int_0^{\pi} d\theta \sqrt{\det \mathbf{g}} = \int_0^{2\pi} d\phi \int_0^{\pi} d\theta R^2 \sin\theta = 4\pi R^2 \qquad (2.27)$$

as expected.

In Euclidean three-space parallel lines of infinite length never cross, but this could not be proved in Euclidean geometry, so it had to be asserted without proof, the *parallel axiom*. The two-sphere belongs to the class of Riemannian curved spaces which are locally flat: a small portion of the surface can be approximated by its tangential plane. Lines in this plane which are parallel locally do, however, cross when drawn far enough, as required for geodesics on the surface of a sphere.

Whether we live in three or more dimensions, and whether our space is flat or curved, is really a physically testable property of space. This was first realized by Carl Friedrich Gauss (1777–1855), who proceeded to investigate it by measuring the angles in a triangle formed by three distant mountain peaks. If space were Euclidean the value would be 180°, but if it were curved the angular sum could be either more or less than 180°. The precision was, however, not good enough to exhibit any disagreement with the Euclidean value .

The deviation of a curved surface from flatness can also be determined from the length of the circumference of a circle. Choose a point P on the surface and draw the locus corresponding to a fixed distance s from that point. If the surface is flat, a plane, the locus is a circle and s is its radius. On a two-sphere of radius R the locus is also a circle, see Fig. 7, but the distance s is measured along a geodesic. The angle subtended by s at the centre of the sphere is s/R, so the radius of the circle is $r = R\sin(s/R)$. Its circumference is then

$$C = 2\pi R \sin(s/R) = 2\pi s \left(1 - \frac{s^2}{6R^2} + ... \right) . \qquad (2.28)$$

Gauss discovered an invariant characterizing the curvature of two-surfaces, the *Gaussian curvature K*. Although K can be given by a completely general formula independent of the coordinate system (see e.g. reference [1]), it is most simply described in an orthogonal system x,y. Let the radius of curvature along the x axis be $R_x(x)$ and along the y axis be $R_y(y)$. Then the Gaussian curvature at the point (x_0,y_0) is

$$K = 1/R_x(x_0)R_y(y_0) . \qquad (2.29)$$

On a two-sphere $R_x = R_y = R$, so $K = R^{-2}$ everywhere. Inserting this into Eq. (2.28) we obtain in the limit of small s

$$K = \frac{3}{\pi}\lim_{s \to 0} \left(\frac{2\pi s - C}{s^3}\right) . \qquad (2.30)$$

This expression is true for any two-surface, and it is in fact the only invariant that can be defined.

The angles in a triangle on a surface with positive curvature like a two-sphere add up to more than 180°, whereas on a saddle surface with negative curvature they add up to less than 180°. This is illustrated in Figs. 8 and 9.

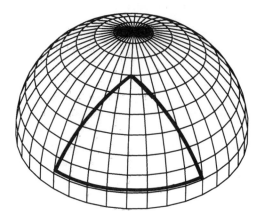

Fig. 8   The angles in a triangle on a surface with positive curvature add up to more than 180°

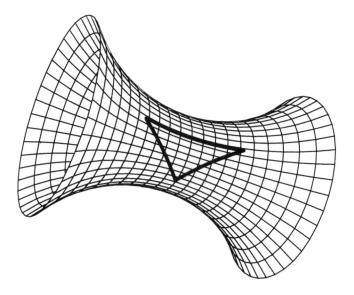

Fig. 9   The angles in a triangle on a surface with negative curvature add up to less than 180°

Returning to the space we inhabit, we manifestly observe that there are three spatial coordinates, so our space must have at least one dimension more than a two-sphere. It is easy to generalize from the curved two-dimensional manifold (surface) (2.22) embedded in three-space to the curved three-dimensional manifold (hypersurface)

$$x^2 + y^2 + z^2 + w^2 = S^2 \tag{2.31}$$

of a three-sphere (hypersphere) embedded in Euclidean four-space with coordinates x,y,z and a fourth fictitious space coordinate w.

Just as the metric (2.23) could be written without explicit use of z, the metric on the three-sphere (2.31) can be written without use of w,

$$dl^2 = dx^2 + dy^2 + dz^2 + \frac{(x dx + y dy + z dz)^2}{S^2 - x^2 - y^2 - z^2} , \qquad (2.32)$$

or in the more convenient spherical coordinates (2.17),

$$dl^2 = \frac{S^2 dR^2}{S^2 - R^2} + R^2 d\theta^2 + R^2 \sin^2\theta \ d\phi^2 . \qquad (2.33)$$

No point is preferred on the manifold (2.31), and hence it can describe a spatially homogeneous and isotropic three-dimensional universe in accord with the cosmological principle.

Another example of a curved Riemannian two-space is the surface of a hyperboloid obtained by changing the sign of S^2 in Eq. (2.22). The geodesics are hyperbolas, the surface is also unbounded, but in contrast to the spherical surface it is infinite in extent. It can also be generalized to a three-dimensional curved hypersurface, a three-hyperboloid, defined by Eq. (2.31) with S^2 replaced by $-S^2$.

The Gaussian curvature of all geodesic three-surfaces in Euclidean four-space is

$$K = k/R^2 , \qquad (2.34)$$

where the *curvature parameter* k can take the values $+1$, 0, -1, corresponding to the three-sphere, flat three-space, and the three-hyperboloid, respectively.

Let us now include the time coordinate t and the curvature parameter k into Eq. (2.33). We also replace the *proper distance* R by the dimensionless ratio $\sigma = R/S$. We then obtain the complete metric derived by Howard Robertson and Arthur Walker in 1934,

$$ds^2 = c^2 dt^2 - dl^2 = c^2 dt^2 - S^2 \left(\frac{d\sigma^2}{1 - k\sigma^2} + \sigma^2 d\theta^2 + \sigma^2 \sin^2\theta \ d\phi^2 \right) . \qquad (2.35)$$

Positive curvature, $k = +1$, obviously corresponds to a metric with the spatial part (2.33), whereas $k = 0$ corresponds to the metric (2.21) of flat Minkowski spacetime. Actually k can take any positive or negative values, but we can always rescale σ to take account of different values in k. Thus all cases are covered in full generality with k restricted to the values $-1, 0, +1$.

If the two-sphere with surface (2.22) and radius S were a balloon expanding with time, $S = S(t)$, points on the surface of the balloon would find their mutual distances scaled by $S(t)$ relative to a time t_0 when the radius was $S(t_0) = 1$. An observer located on any one point would see all the other points recede radially. This is exactly how we see distant galaxies, except that we are not on a two-sphere but rather on a spatially curved three-surface with cosmic scale factor $S(t)$ and curvature parameter k unknown.

Thus if the Universe is homogeneous and isotropic at one time and has the *Robertson–Walker metric* (2.35), then *it will always remain homogeneous and isotropic*, because a galaxy at the point (σ,θ,ϕ) will always remain at that point, only the scale of spatial distances $S(t)$ changing with time. The displacements will be $d\sigma = d\theta = d\phi = 0$ and the metric equation will reduce to

$$ds^2 = c^2 dt^2 . \tag{2.36}$$

For this reason one calls such an expanding frame a *comoving frame* and the coordinates (σ,θ,ϕ) *comoving coordinates*.

An observer at rest in the comoving frame is called a *fundamental observer*. If the Universe appears homogeneous to him, it must also be isotropic. But another observer located at the same point and in relative motion with respect to the fundamental observer does not see the Universe as isotropic. Thus the comoving frame is really a preferred frame, and a very convenient one, as we shall see later in conjunction with the cosmic background radiation. Let us note here that a fundamental observer may find that not all astronomical bodies recede radially, but that some exhibit peculiar motion in other directions.

Let us find the distance l from our Galaxy, chosen as the origin, to another galaxy at comoving coordinates $(\sigma,0,0)$. The geodesic lies on a hypersurface in the four-space of the Robertson–Walker metric with the spatial distance $dl \equiv |\,\mathbf{dl}|$ given by Eq. (2.35). Integrating dl from 0 to l we find

$$l = S(t) \int_0^\sigma \frac{d\sigma}{\sqrt{1 - k\sigma^2}} \tag{2.37}$$

which depends on the curvature parameter k. For flat space $k = 0$ we find the expected result $l = S\sigma$. Note that σ grows with l without limit. This shows that flat space is unbounded.

In a universe with curvature $k = +1$ the integral in (2.37) yields

$$l = S\sin^{-1}\sigma, \quad \text{or} \quad \sigma = \sin(l/S) .$$

As the distance l increases from 0 to $\frac{1}{2}\pi S$, σ also increases from 0 to its maximum value 1. However, when l increases from $\frac{1}{2}\pi S$ to πS, σ decreases back to 0, and we are back at the origin. Thus, travelling a distance $l = \pi S$ through the curved three-space brings us back to the point of departure. In this sense a universe with positive curvature is bounded and closed.

Similarly, the area of a three-sphere centred at the origin and going through the second galaxy at comoving distance σ is

$$A = 4\pi S^2 \sigma^2 = 4\pi S^2 \sin^2(l/S) . \tag{2.38}$$

Clearly, A goes through a maximum when $l = \frac{1}{2}\pi S$, and decreases back to 0 when l reaches πS. Note that $A/4$ equals the surface of the circle formed by intersecting a two-sphere of radius S with a horizontal plane, as shown in Fig. 7. The intersection with an equatorial plane results in the circle with maximal surface, $A/4 = \pi S^2$, all other intersections making smaller circles. A plane tangential at either pole has no intersection, thus the corresponding 'circle' has zero surface.

The volume of the three-sphere (2.31) can then be written in analogy with Eq. (2.27),

$$V = \int_0^{2\pi} d\phi \int_0^{\pi} d\theta \int_0^1 d\sigma \sqrt{\det \mathbf{g}_{RW}} , \qquad (2.39)$$

where the determinant of the spatial metric matrix $\mathbf{g}_{RW}$ is now

$$\det \mathbf{g}_{RW} = S^6 \frac{\sigma^4}{1 - \sigma^2} \sin^2\theta . \qquad (2.40)$$

Thus one finds the volume of the three-sphere

$$V = 2\pi^2 S^3 . \qquad (2.41)$$

The hyperbolic case is different. Setting $k = -1$ in Eq. (2.37) we find

$$l = S\sinh^{-1}\sigma, \quad \text{or} \quad \sigma = \sinh(l/S) . \qquad (2.42)$$

Clearly this space is unbounded because σ grows indefinitely with l. The area of the three-hyperboloid through the galaxy at σ is

$$A = 4\pi S^2 \sigma^2 = 4\pi S^2 \sinh^2(l/S) . \qquad (2.43)$$

A universe of this kind is therefore unbounded and open.

Let us differentiate l in Eq. (2.37) with respect to time, noting that σ is a constant since it is a comoving coordinate. Then we obtain the recession velocity v of the galaxy at distance l,

$$v = \dot{l} = \dot{S}(t) \int_0^{\sigma} \frac{d\sigma}{\sqrt{1 - k\sigma^2}} = \frac{\dot{S}(t)}{S(t)} l. \qquad (2.44)$$

Thus the recession velocity is proportional to distance, and Hubble's law emerges in a form more general than Eq. (1.28):

$$H(t) = \frac{\dot{S}(t)}{S(t)} . \qquad (2.45)$$

In Eq. (1.24) we defined the Hubble radius r_H as the distance reached in one Hubble time, τ_H, by a light signal propagating along a straight line in flat, static space. In a spacetime described by the Robertson–Walker metric the light signal propagates along the geodesic $ds^2 = 0$. Let us define the *particle horizon* or the *object horizon* σ_p as the largest comoving distance from which a light signal could have reached us if it was emitted at time $t = 0$. Thus it delimits the causally connected part of the Universe an observer can see at a given time t. Taking our coordinates to be $\sigma = 0$, t_0, then σ_p is determined from Eq. (2.35),

$$\int_0^{\sigma_p} \frac{d\sigma}{\sqrt{1 - k\sigma^2}} = c \int_0^{t_0} \frac{dt}{S(t)} \equiv \chi , \qquad (2.46)$$

where we have introduced the abbreviation χ for the time integral. Solving for σ_p,

$$\sigma_p = \sin\chi \text{ for } k = +1 ,$$
$$\chi \text{ for } k = 0 ,$$
$$\sinh\chi \text{ for } k = -1 . \qquad (2.47)$$

A particle horizon exists if the origin of time at $t = 0$ is in the finite past. Clearly the value of σ_p depends sensitively on the behaviour of the scale of the Universe at that time, $S(t_p)$.

If $k \geq 0$, the proper distance to the particle horizon at time t is

$$r_p = S(t) \, \sigma_p . \qquad (2.48)$$

Note that r_p equals the Hubble radius $r_H = c/H_0$ when $k = 0$ and the scale is a constant, $S(t) = S$. When $k = -1$ the Universe is open, and r_p cannot be interpreted as a measure of its size.

In an analogous way we can define the comoving distance σ_e to the *event horizon*, defined as the most distant present event from which a world line can ever reach our world line. By 'ever' is meant a finite future time, t_{max}:

$$\int_0^{\sigma_e} \frac{d\sigma}{\sqrt{1 - k\sigma^2}} = c \int_{t_0}^{t_{max}} \frac{dt}{S(t)} . \qquad (2.49)$$

The particle horizon $\sigma_p(t_0)$ at time t_0 lies on our past light cone, but with time our particle horizon will broaden to $\sigma_p(t > t_0)$ so that the light cone at t_0 will move inside the light cone at $t > t_0$, see Fig. 3. The event horizon at this moment can only be specified given the time distance to the ultimate future, t_{max}. Only at t_{max} will our past light cone encompass the present event horizon. Thus the event horizon is our ultimate particle horizon. The integrands in Eqs. (2.46) and (2.49) are obviously the same, only the integration limits showing that the two horizons correspond to different viewpoints.

The proper distance to the particle horizon is $r_p = S\sigma_p$. Comoving bodies at the particle horizon recede with velocity $v = Hr_p$, but the particle horizon itself recedes even faster, with velocity

$$\dot{r}_p = \dot{S}\sigma_p + S\frac{c}{S} = Hr_p + c . \qquad (2.50)$$

Thus when the particle horizon grows with time, bodies which were at spacelike distances at earlier times enter into the light cone.

One way of deciding whether the Universe is closed or open is to look for clues that constellations of stars recur. If the Universe is closed and not too large we could see our own Galaxy at some distance since the line of sight is a closed loop. Alternatively, one could look for the same astronomical object in two opposite directions. Fang and Liu [3] have undertaken this by looking for identical quasars within $\pm 2°$ of opposite directions, and within $\pm 5\%$ of identical distances (in z). From a catalogue of 47 quasars compiled in 1988, all at $z > 1$, they found no identical pair. This negative result is quite interesting because it sets a lower limit to the radius of the Universe if it is closed: $R \gtrsim 400 \ h^{-1}$Mpc. This is very small indeed in comparison with the Hubble radius r_H defined in Eq. (1.24), so it certainly does not contradict the possibility of a closed Universe.

2.3 The Principle of Covariance

A more general way to write metric equations is

$$ds^2 = g_{\mu\nu}dx^\mu dx^\nu , \qquad (2.51)$$

where summation over μ and ν is implied. Greek letters are used when all the four coordinates x^μ, $\mu = 0, 1, 2, 3$, are implied; roman letters when only the spatial components x^i, $i = 1, 2, 3$, are implied. The time coordinate is $x_0 = ct$ and the spatial coordinates may correspond to, for instance, the Cartesian x, y, z, or the spherical r, θ, ϕ. The quantities $g_{\mu\nu}$ are the components of a 4×4 matrix. In the language of general relativity this matrix is a tensor called the *metric tensor*. In tensor calculus there are rules for when to write coordinate indices as subscripts and when as superscripts. We shall not elaborate on these rules here.

The geometry of curved spaces had been studied in the nineteenth century by Gauss, Riemann and others. The key quantities are *tensors* which Einstein learned from his friend Marcel Grossman. In general, a tensor is a quantity with two or more indices running over the dimensions of a manifold. Thus if the manifold is d-dimensional and the tensor has r indices, it is a *rank r* tensor with d^r components. Clearly if $r = 1$, the corresponding quantity is a vector, and if $r = 0$, it is a scalar. An example of a tensor is the assembly of the n^2 components $A_\mu B_\nu$ formed as the products of the n components of the vector A_μ with the n components of the vector B_ν. Another well-known example of a rank 2 tensor is the Kronecker delta with components $\delta_{\mu\nu}$.

Orthogonal coordinate systems have diagonal metric tensors, and this is all what we shall encounter. Thus the components of the Robertson–Walker metric (2.35) can be written as a diagonal matrix with non-vanishing elements

$$g_{00} = 1, \quad g_{11} = -\frac{S^2}{1 - k\sigma^2}, \quad g_{22} = -S^2\sigma^2, \quad g_{33} = -S^2\sigma^2\sin^2\theta. \qquad (2.52)$$

We already made use of this metric matrix in Eq. (2.40) where we called it $\mathbf{g}_{RW}$.

For the Minkowski metric tensor (1.4) one uses the notation $\eta_{\mu\nu}$ with the diagonal components

$$\eta_{00} = 1, \quad \eta_{jj} = -1, \quad j = 1,2,3, \qquad (2.53)$$

all non-diagonal components vanishing.

In four-dimensional spacetime all spatial three-vectors have to acquire a zero-component just like the distance four-vector ds with components cdt, dx, dy, dz. Thus the four-momentum P with spatial components p_x, p_y, p_z has the energy E/c as zero-component so that E and $\mathbf{p}$ become two aspects of the same entity, $P = (E/c, \mathbf{p})$.

The scalar product P^2 is an invariant related to the mass,

$$P^2 = \eta_{\mu\nu} P_\mu P_\nu = \frac{E^2}{c^2} - p^2 = m^2 c^2, \tag{2.54}$$

where $p^2 \equiv |\mathbf{p}|^2$. For a particle at rest, $\mathbf{p}=0$, Einstein's famous relation follows,

$$E = mc^2 . \tag{2.55}$$

For a massless particle like the photon it follows that the energy equals the three-momentum times c^2.

Newton's second law in its non-relativistic form

$$\mathbf{F} = m\mathbf{a} = m\dot{\mathbf{v}} = \dot{\mathbf{p}} \tag{2.56}$$

is replaced by the relativistic expression

$$F = \frac{dP}{d\tau} = \gamma \frac{dP}{dt} , \tag{2.57}$$

where F is the force four-vector

$$F = \left(\frac{dE}{cd\tau} , \frac{d\mathbf{p}}{d\tau} \right) = \gamma \left(\frac{dE}{cdt} , \frac{d\mathbf{p}}{dt} \right) ,$$

and γ is the Lorentz factor $(1 - v^2/c^2)^{-\frac{1}{2}}$.

Although Newton's second law (2.57) is invariant under special relativity in any inertial frame, it is not invariant in accelerated frames. Since this law explicitly involves acceleration, special relativity has to be generalized somehow, so that observers in accelerated frames can agree on the value of acceleration. Thus the next necessary step is to search for quantities which remain invariant under an arbitrary acceleration and to formulate the laws of physics in terms of these. Such a formulation is called *generally covariant*.

Since we live in a curved spacetime described by the Robertson–Walker metric, the approach to general covariance is to require that the invariants are the same in curved spacetimes. We are then turning to tensors which are mathematical objects which have the desired properties.

Physical laws formulated in terms of tensors retain their form on all curved surfaces. Vectors are tensors, so vector equations may already be covariant. However, dynamical laws contain many other quantities which are not tensors, in particular spacetime derivatives such as $d/d\tau$ in Eq. (2.57). Spacetime derivatives are not invariants because they imply transports ds along some curve and that makes them coordinate-dependent. Therefore we have to start by redefining derivatives and replace them by new *covariant derivatives* which are tensor quantities.

To make the spacetime derivative of a vector generally covariant one has to take into account that the direction of a parallel-transported vector changes in terms of the local coordinates along the curve, see Fig. 10. The change is certainly some function of the spacetime derivatives of the curved space which is described by the metric tensor.

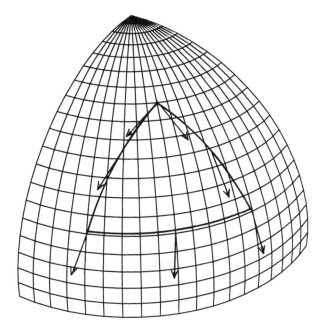

Fig. 10   Parallel transport of a vector around a closed contour on a curved surface

We are not going to carry through the explicit derivation of the expression for the change in the derivative of a vector entering the expression for the covariant derivative operator $D/D\tau$ (for more details, see, for example, references [1], [2]). The resulting covariant derivative of the momentum four-vector P^μ with respect to τ appearing in Eq. (2.57) is

$$F^\mu = \frac{DP^\mu}{D\tau} \equiv \frac{dP^\mu}{d\tau} + \Gamma^\mu_{\sigma\nu} P^\sigma \frac{dx^\nu}{d\tau} \, , \tag{2.58}$$

where summation from 0 to 3 over repeated indices is implied. The second term contains the changes this vector undergoes when it is parallel-transported the infinitesimal distance $cd\tau$. The quantities $\Gamma^\mu_{\sigma\nu}$ called *affine connections* are readily derivable functions of the derivatives of the $g_{\mu\nu}$ in the curved spacetime (they are not tensors). Their form is

$$\Gamma^\mu_{\sigma\nu} = \tfrac{1}{2} g^{\mu\rho} \left(\frac{\partial g_{\sigma\rho}}{\partial x^\nu} + \frac{\partial g_{\nu\rho}}{\partial x^\sigma} - \frac{\partial g_{\sigma\nu}}{\partial x^\rho} \right) \, . \tag{2.59}$$

With this definition Newton's second law has been made generally covariant.

The path of a test body in free fall, its geodesic, follows from Eq. (2.58) by requiring that no forces act on the body, $F^\mu = 0$. Making the replacement

$$P^\mu = m \frac{dx^\mu}{d\tau} \, ,$$

the equation of the geodesic can be written

$$\frac{d^2x^\mu}{d\tau^2} + \Gamma^\mu_{\sigma\nu} \frac{dx^\sigma}{d\tau} \frac{dx^\nu}{d\tau} = 0 . \tag{2.60}$$

In an inertial frame the metric is flat, the metric tensor is a constant at every point x,

$$g_{\mu\nu}(x) = \eta_{\mu\nu} ,$$

and thus the spacetime derivatives of the metric tensor vanish,

$$\frac{\partial g_{\mu\nu}(x)}{\partial x^\rho} = 0 . \tag{2.61}$$

It then follows from Eq. (2.59) that also the affine connections vanish, and the covariant derivatives equal the simple spacetime derivatives.

Going from an inertial frame at x to an accelerated frame at $x + \Delta x$ the expressions for $g_{\mu\nu}(x)$ and its derivatives around x can be obtained as the Taylor expansions

$$g_{\mu\nu}(x + \Delta x) = \eta_{\mu\nu} + \frac{1}{2} \frac{\partial^2 g_{\mu\nu}(x)}{\partial x^\rho \partial x^\sigma} \Delta x^\rho \Delta x^\sigma + ...$$

and

$$\frac{\partial g_{\mu\nu}(x + \Delta x)}{\partial x^\rho} = \frac{\partial^2 g_{\mu\nu}(x)}{\partial x^\rho \partial x^\sigma} \Delta x^\sigma + ...$$

The description of a curved spacetime thus involves second derivatives of $g_{\mu\nu}$, at least. We have already introduced one non-covariant curvature parameter, the Gaussian curvature K defined on a two-dimensional surface. This also depends on the second derivatives of $g_{\mu\nu}$.

In a higher-dimensional spacetime, curvature has to be defined in terms of more than just one parameter K. It turns out that curvature is most conveniently defined in terms of the fourth rank *Riemann tensors* $R^\alpha_{\beta\gamma\delta}$ and $R_{\sigma\beta\gamma\delta}$, which are functions of second derivatives of $g_{\mu\nu}$ (see e.g. references [1], [2]). In four-space these tensors have 256 components, but most of them vanish or are not independent because of several symmetries and antisymmetries in the indices. Moreover, an observer at rest in the comoving Robertson–Walker frame will only need to refer to spatial curvature. Using the indices $i, j, k, l = 1, 2, 3$ we can write the Riemann tensor

$$R_{ijkl} = \frac{k}{S^2(t)} \ (g_{ik}g_{jl} - g_{il}g_{kj}) . \tag{2.62}$$

In an n-manifold R_{ijkl} has only $n^2(n^2 - 1)/12$ nonvanishing components, thus six in the spatial three-space of the Robertson–Walker metric. On the two-sphere there is only one component which essentially is the Gaussian curvature K.

Another important tool related to curvature is the second rank *Ricci tensor* $R_{\beta\gamma}$, obtained from the Riemann tensor by a summing operation over repeated indices called *contraction*,

$$R_{\beta\gamma} = R^{\alpha}_{\beta\gamma\alpha} = \delta^{\delta}_{\alpha} R^{\alpha}_{\beta\gamma\delta} = g^{\alpha\delta} R^{\alpha}_{\beta\gamma\delta} = \frac{1}{g_{\alpha\delta}} R^{\alpha}_{\beta\gamma\delta} \ . \tag{2.63}$$

This n^2-component tensor is symmetric in the two indices, so it has only $\frac{1}{2}n(n+1)$ independent components. In four-space the ten components of the Ricci tensor lead to Einstein's system of ten gravitational equations as we shall see later.

Finally we may sum over the two indices of the Ricci tensor to obtain the *Ricci scalar* R,

$$R = g^{\beta\gamma} R_{\beta\gamma} = \frac{1}{g_{\beta\gamma}} R_{\beta\gamma} \ . \tag{2.64}$$

This scalar we will also need later.

2.4 The Principle of Equivalence

Newton's law of gravitation, Eq. (2.1), runs into serious conflict with special relativity in three different ways. Firstly, there is no obvious way of rewriting it in terms of invariants, since it only contains scalars. Secondly, it has no explicit time dependence, so gravitational effects are propagated instantaneously to every location in the Universe (in fact, also infinitely far outside the horizon of the Universe!).

Thirdly, the *gravitating mass m_G* appearing in Eq. (2.1) is totally independent of the *inert mass m* appearing in Newton's second law (2.56), as we already noted, yet for unknown reasons both masses appear to be equal to high precision. Clearly a theory is needed establishing a formal link between them. Mach thought that the inert mass of a body was somehow linked to the gravitational mass of the whole Universe. Einstein, who was strongly influenced by the ideas of Mach, called this *Mach's principle*. In his early work on general relativity he considered it as one of the basic, underlying principles together with the principles of equivalence and covariance, but in the end he did not refer to it anymore.

Facing the above shortcomings of Newtonian mechanics and the limitations of special relativity Einstein set out on a long and tedious search for a better law of gravitation, yet one that would reduce to Newton's law in some limit, of the order of the precision of planetary mechanics.

Consider the elevator in Fig. 11 moving vertically in a tall tower (it is easy to imagine an elevator to be at rest with respect to an outside observer fixed to the tower, whereas the more 'modern' example of a space craft is not at rest when we observe it to be geostationary). A passenger in the elevator testing the law of gravitation would find that objects dropped to the floor acquire the usual gravitational acceleration g when the elevator stands still, or moves with constant speed. However, when the outside observer notes that the elevator is accelerating upwards, tests inside the elevator reveal that the objects acquire an acceleration larger than g, and vice versa when the elevator is accelerating downwards. In the limit of free fall (unpleasant to the passenger) the objects appear weightless corresponding to zero acceleration.

Let us now replace the elevator by a space craft with the engines turned off, located at some neutral point in space where all gravitational pulls cancel or are negligible

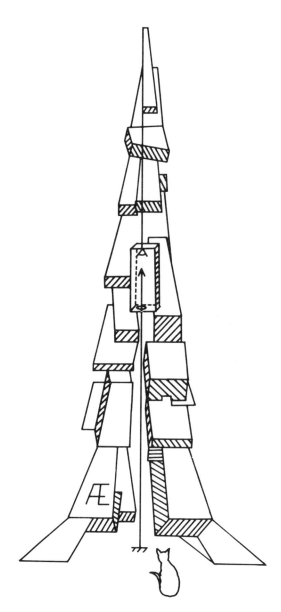

Fig. 11 The Einstein elevator mounted in a non-Euclidean tower. The observer is seen in the foreground

(a good place is the *Lagrange point* where the terrestrial and solar gravitations cancel). All objects as well as the pilot would appear weightless.

Now turning on the engines by remote radio control, the space craft could be accelerated upwards so that objects on board would acquire an acceleration g toward the floor. The pilot would then rightly conclude that *gravitational pull and local acceleration are equivalent* and indistinguishable if no outside information is available and if $m = m_G$. This conclusion forms the important *principle of equivalence* which states that

*to an observer in free fall in a gravitational field the results of all local
experiments are completely independent of the magnitude of the field.*

A passenger in the elevator measuring g could well decide that Earth's gravitation
actually does not exist, but that the elevator is accelerating radially outwards from Earth.
This interpretation would come in conflict with that of another observer on the opposite
side of Earth whose frame would accelerate in the opposite direction, so how could
Earth's gravitation generally be replaced by acceleration?

The answer is that the principle of equivalence is only true *locally*, in a suitably
small elevator or space craft where the curved spacetime also can be matched by flat
Minkowski spacetime. In the gravitational field of Earth the gravitational acceleration is
directed toward its centre. Thus the two test bodies in Fig. 12 with a horizontal separation
do not actually fall along parallel lines, but along different radii so that their separation
decreases with time. This phenomenon is called the *tidal effect*, or sometimes the tidal
force since the test bodies move as if an attractive exchange force acted upon them.

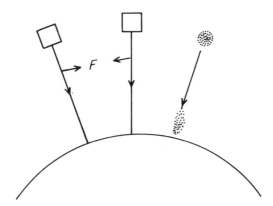

Fig. 12 Tidal force F acting between two test bodies falling freely towards the surface of a
gravitating body. On the right a spherical cluster of small bodies is seen to become ellipsoidal on
approaching the body

An interesting example is offered by a sphere of freely falling particles. Since the
strength of the gravitational field increases in the direction of fall, the particles in the
front of the sphere will fall faster than those in the rear. At the same time the lateral
cross-section of the sphere will shrink due to the tidal effect. As a result, the sphere
will be *focused* into an ellipsoid with the same volume. This effect is responsible for the
gravitational breakup of very nearby massive stars.

If the tidal effect is too small to be observable, the laboratory can be considered local.
On a larger scale the gravitational field is clearly quite inhomogeneous, so if we make use
of the principle of equivalence to replace this field everywhere by local flat frames, we
get a patchwork of frames which describe a curved spacetime. Since the inhomogeneity
of the field is caused by the inhomogeneous distribution of gravitating matter, Einstein
realized that the spacetime we live in had to be curved, and the curvature had to be
related to the distribution of matter.

Let us return once more to the passenger in the Einstein elevator for a demonstration of the relation between gravitation and the curvature of spacetime. Let the elevator be in free fall; the passenger would consider that no gravitational field is present. Standing by one wall and shining a pocket lamp horizontally across the elevator, she sees that light travels in a straight path, a geodesic in flat spacetime. This is illustrated in Fig. 13. Thus she concludes that in the absence of a gravitational field spacetime is flat.

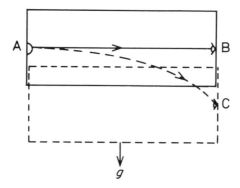

Fig. 13   A pocket lamp at A in the Einstein elevator is shining horizontally on a point B. However, an outside observer who sees that the elevator is falling freely with acceleration g concludes that the light ray follows the dashed curve to point C

However, the outside observer in the tower sees that the elevator has accelerated while the light front travels across the elevator, and so with respect to the fixed frame of the tower he notices that the light front follows a curved path, as shown in Fig. 13. He also sees that the elevator is falling in the gravitational field of Earth, and so he would conclude that light feels gravitation as if it had mass. He could also phrase it differently: light follows a geodesic, and since this light path is curved it must imply that spacetime is curved in the presence of a gravitational field.

When the passenger shines monochromatic light of frequency ν vertically, it reaches the roof at height d in time d/c. In the same time the outside observer notices that the elevator has accelerated from, say $v = 0$ to gd/c, where g is the gravitational accceleration on Earth, so that the colour of the light has redshifted by the fraction

$$\frac{\Delta\nu}{\nu} \approx \frac{v}{c} = \frac{gd}{c^2} = \frac{GMd}{r^2c^2} \; . \tag{2.65}$$

Thus the photons have lost energy ΔE by climbing the distance d against Earth's gravitational field,

$$\Delta E = h\Delta\nu = -\frac{gdh\nu}{c^2} \; , \tag{2.66}$$

where h is the *Planck constant*. Max Planck (1858–1947) was the inventor of the energy quantum; this led to the discovery and development of *quantum mechanics*.

If the pocket lamp had been shining electrons of mass m, they would have lost kinetic energy

$$\Delta E = -gmd \tag{2.67}$$

climbing up the distance d. Combining Eqs. (2.66) and (2.67) we see that the photons appear to possess mass

$$m = \frac{h\nu}{c^2} \ . \tag{2.68}$$

Eq. (2.65) clearly shows that light emerging from a star with mass M is redshifted in proportion to M. Thus part of the redshift observed is due to this gravitational effect. From this we can anticipate the existence of stars with such a large mass that their gravitational field effectively prohibits radiation to leave. These are the *black holes* to which we shall return later.

2.5 Einstein's Theory of Gravitation

Realizing that the space we live in was not flat, except locally and approximately, Einstein proceeded to combine the principle of equivalence with the requirement of general covariance. The inhomogeneous gravitational field near a massive body being *equivalent* to (a patchwork of flat frames describing) a curved spacetime, the laws of Nature (such as the law of gravitation) have to be described by *generally covariant* tensor equations. Thus the law of gravitation has to be a covariant relation between mass density and curvature.

The starting point is Newton's law of gravitation, because this has to be true anyway in the limit of very weak fields. From Eq. (2.2), the gravitational force experienced by a unit mass at distance r from a body of mass M and density ρ is a vector in three-space

$$\ddot{\mathbf{r}} = \mathbf{F} = \frac{GM}{\mathbf{r}^2}$$

with components ($i = 1,2,3$)

$$\frac{\mathrm{d}^2 x^i}{\mathrm{d}t^2} = F^i = \frac{GMx^i}{r^3} \ . \tag{2.69}$$

Let us define a scalar *gravitational potential* ϕ by

$$\frac{\partial \phi}{\partial x^i} = -F^i \ .$$

This can be written more compactly

$$\nabla \phi = -\mathbf{F} \ . \tag{2.70}$$

Integrating the flux of the force **F** through a spherical surface surrounding M and using Stokes's theorem, one can show that the potential ϕ obeys Poisson's equation

$$\nabla^2\phi = 4\pi G\rho \ . \tag{2.71}$$

Let us next turn to the relativistic equation of motion (2.60). In the limit of weak and slowly varying fields for which all time derivatives of $g_{\mu\nu}$ vanish and the (spatial) velocity components $dx^i/d\tau$ are negligible compared to $dx^0/d\tau = cdt/d\tau$, Eq. (2.60) reduces to

$$\frac{d^2x^\mu}{d\tau^2} + c^2\Gamma^\mu_{00}\left(\frac{dt}{d\tau}\right)^2 = 0 \ . \tag{2.72}$$

From Eq. (2.58) these components of the affine connection are

$$\Gamma^\mu_{00} = -\tfrac{1}{2}g^{\mu\rho}\frac{\partial g_{00}}{\partial x^\rho} \ ,$$

where g_{00} is the time–time component of $g_{\mu\nu}$.

In a weak static field the metric is almost that of flat spacetime, so we can approximate $g_{\mu\nu}$ by

$$g_{\mu\nu} = \eta_{\mu\nu} + h_{\mu\nu} \ ,$$

where $h_{\mu\nu}$ is a small increment to $\eta_{\mu\nu}$. To lowest order in $h_{\mu\nu}$ we can then write

$$\Gamma^\mu_{00} = -\tfrac{1}{2}\eta^{\mu\rho}\frac{\partial h_{00}}{\partial x^\rho} \ . \tag{2.73}$$

Inserting this expression into Eq. (2.72) the equations of motion become

$$\frac{d^2\mathbf{x}}{d\tau^2} = -\tfrac{1}{2}(dt/d\tau)^2c^2\nabla h_{00} \tag{2.74}$$

$$\frac{d^2t}{d\tau^2} = 0 \ . \tag{2.75}$$

Dividing Eq. (2.74) by $(dt/d\tau)^2$ and setting it equal to 1 we obtain

$$\frac{d^2\mathbf{x}}{dt^2} = -\tfrac{1}{2}c^2\nabla h_{00} \ . \tag{2.76}$$

Comparing this with the Newtonian equation of motion (2.69) in the x^i direction we obtain the value of the time–time component of $h_{\mu\nu}$

$$h_{00} = 2\frac{\phi}{c^2} \ ,$$

from which follows that

$$g_{00} = 1 + 2\frac{\phi}{c^2} = 1 - \frac{2GM}{c^2 r} \ . \tag{2.77}$$

Let us now turn to the distribution of matter in the Universe. Suppose that matter can be considered continuously distributed as in an *ideal fluid*. The energy density, pressure, and shear of a fluid of non-relativistic matter are compactly described by the *stress-energy tensor* $T_{\mu\nu}$ with the following components:

(i) The time–time component T_{00} is the energy density ρc^2 which includes the rest mass as well as internal and kinetic energies.
(ii) The diagonal space–space components T_{ii} are the pressure components in the i direction p^i, or the momentum components per unit area.
(iii) The time–space components cT_{0i} are the energy flow per unit area in the i direction.
(iv) The space–time components cT_{i0} are the momentum densities in the i direction.
(v) The non-diagonal space–space components T_{ij} are the shear of the pressure component p^i in the j direction.

It is important to note that the stress-energy tensor is of rank 2 and symmetric, thus it has 10 independent components in four-space. However, a comoving observer in the Robertson–Walker spacetime, who follows the motion of the fluid, sees no time–space or space–time components. Moreover, we can invoke the cosmological principle to neglect the anisotropic non-diagonal space–space components. Thus the stress-energy tensor can be cast into purely diagonal form,

$$T_{\mu\mu} = (p + \rho c^2)\frac{v_\mu v_\mu}{c^2} - p g_{\mu\mu}. \tag{2.78}$$

In particular, the time–time component T_{00} is ρc^2. The conservation of energy and three-momentum, or equivalently the conservation of four-momentum can be written

$$\frac{DT_{\mu\nu}}{Dx_\nu} = 0 \ . \tag{2.79}$$

Thus the stress-energy tensor is divergence-free.

Taking $T_{\mu\nu}$ to describe relativistic matter, one has to pay attention to its Lorentz transformation properties which differ from the classical case. Under Lorentz transformations the different components of a tensor do not remain unchanged, but become dependent on each other. Thus the physics embodied by $T_{\mu\nu}$ also differs: the gravitational field does not depend on mass densities alone, but also on pressure. All the components of $T_{\mu\nu}$ are therefore responsible for warping the spacetime .

We can now put several things together: replacing ρ in the field equation (2.71) by T_{00}/c^2, and substituting ϕ from Eq. (2.77) we obtain a field equation for weak static fields generated by non-relativistic matter,

$$\nabla^2 g_{00} = \frac{8\pi G}{c^4} T_{00} \ . \tag{2.80}$$

Let us now assume with Einstein that the right-hand side could describe the source term of a relativistic field equation of gravitation if we made it generally covariant. This suggests replacing T_{00} by $T_{\mu\nu}$. In a matter-dominated universe where the gravitational field is produced by massive stars, and where the pressure between stars is negligible, the only component of importance is then T_{00}.

The left-hand side of Eq. (2.80) is not covariant, but it does contain second derivatives of the metric, albeit of only one component. Thus it is already related to curvature. The next step would be to replace $\nabla^2 g_{00}$ by a tensor matching the properties of $T_{\mu\nu}$ on the right-hand side:

(i) It should be of rank two.
(ii) It should be related to the Riemann curvature tensor $R_{\sigma\beta\gamma\delta}$. We have already found a candidate in the Ricci tensor $R_{\mu\nu}$ in Eq. (2.63).
(iii) It should be symmetric in the two indices. This is true for the Ricci tensor.
(iv) It should be divergenceless in the sense of covariant differentiation. This is not true for the Ricci tensor, but a divergenceless combination can be formed with the Ricci scalar R in Eq. (2.64),

$$ G_{\mu\nu} = R_{\mu\nu} - \tfrac{1}{2} g_{\mu\nu} R \; . \tag{2.81} $$

$G_{\mu\nu}$ is called the *Einstein tensor*. It contains only terms which are either quadratic in the first derivatives of the metric tensor or linear in the second derivatives.

Thus we arrive at Einstein's covariant formula for the law of gravitation,

$$ G_{\mu\nu} = \frac{8\pi G}{c^4} T_{\mu\nu} \; . \tag{2.82} $$

For weak stationary fields produced by non-relativistic matter G_{00} indeed reduces to $\nabla^2 g_{00}$. The Einstein tensor vanishes for flat spacetime and in the absence of matter and pressure, as it should.

Thus the problems encountered by Newtonian mechanics and discussed at the end of Section 2.1 have been resolved in Einstein's theory. The recession velocities of distant galaxies do not exceed the speed of light, and effects of gravitational potentials are not felt instantly, because the theory is relativistic. The discontinuity of homogeneity and isotropy at the boundary of the Newtonian universe has also disappeared. In general relativity spacetime is generated by matter, thus spacetime itself ceases to exist where matter does not exist, so there cannot be any boundary between a homogeneous universe and an empty outside spacetime.

Problems

1. Prove Newton's theorem that the gravitational force at a radial distance R from the centre of a spherical distribution of matter acts as if all the mass inside R were concentrated at a single point at the centre. Show also that if the spherical distribution of matter extents beyond R, the force due to the mass outside R vanishes[4].

2. Suppose that the Universe is decelerating with $q = 1/2$ constantly. We observe two galaxies in opposite directions, both at proper distance r. What is the maximum separation between the galaxies at which they are still causally connected? Express your result as a fraction of distance to r_p. What is the observer's particle horizon?

3. Show that the Hubble distance $r_H = c/H$ recedes with radial velocity

$$\dot{r}_H = c(1 + q) \; . \tag{2.83}$$

What function of time should the scale factor $S(t)$ be in order for r_H to stay constant in time?

4. Express the Hubble parameter $H(t)$ as a function of the deceleration parameter q_0 and time t.

5. Is the sphere defined by the Hubble radius r_H inside or outside the particle horizon ?

6. Calculate whether the following spacetime intervals from the origin are spacelike, timelike, or lightlike: (1, 3, 0, 0); (3, 3, 0, 0); (3, -3, 0, 0); (0, 3, 0, 0); (3, 1, 0, 0)[1].

7. The supernova 1987A explosion in the large Magellanic cloud 170 000 light-years from Earth produced a burst of antineutrinos $\bar{\nu}_e$ which were observed in terrestrial detectors. If the antineutrinos are massive, their velocity would depend on their mass as well as their energy. What is the proper time interval between the emission, assumed instantaneous, and the arrival on the Earth? Show that in the limit of vanishing mass the proper time interval is zero. What information can be derived about the antineutrino mass from the observation that the energies of the antineutrinos ranged from 7 to 11 MeV, and the arrival times showed a dispersion of 7s?

8. A coded light signal is sent from Earth in a particular direction. Our descendants receive this signal from the opposite direction in 1 Gyr. Assuming that the Universe has a closed Robertson–Walker metric, what is its curvature ?

9. Calculate the gravitational redshift in wavelength for the 769.9 nm potassium line emitted from the Sun's surface[1].

10. What is the metric equation of an isotropic homogeneous three-dimensional space with curvature -10^{-2} m^{-2}? What is the area of a spherical surface whose measured distance from the central point is 10 m? [1]

References

1. I. R. Kenyon, *General Relativity*, Oxford University Press, Oxford, 1990.
2. P. J. E. Peebles, *Principles of Physical Cosmology*. Princeton University Press, Princeton, New Jersey, 1993.
3. L. Z. Fang and Y. L. Liu, *Modern Physics Letters*, **A3** (1988) 1221.
4. F. H. Shu, *The Physical Universe*, University Science Books, Mill Valley, CA, 1982.

3 Cosmological Models

If the Einstein equations (2.82) were difficult to derive, it may be even more difficult to find solutions to this system of 10 coupled non-linear differential equations. A particularly simple case, however, is a single star at rest in infinite empty space. Clearly this approximates to the situation of a single star far away from the gravitational influence of other bodies, and it can be tested within the solar system. This is the Schwarzschild solution to the Einstein equations which we shall meet in Section 3.1. The most fascinating consequence of it is black holes.

In Section 3.2 we turn to the 'Standard' or FRW model of cosmology which is based on Friedman's equations and the Robertson–Walker metric, and which takes both energy density and pressure to be functions of time in a Copernican universe.

In Section 3.3 we describe the de Sitter models which do not describe the Universe at large as we see it now, but which has important applications to black holes and to the very early universe.

The predictions of theoretical models can be tested in the physical world by measuring the values of theoretical parameters. We need only a few parameters to test the standard model. In Section 3.4 we establish which they are and how they are related to each other. Later on we shall test them in situations where they give important information about the possible models of the Universe.

3.1 The Schwarzschild Solution and Black Holes

In this Section we shall study the properties of the gravitational field of a spherical star of mass M in isolation from all other gravitational influences. Suppose that we want to measure time t and radial elevation r in the vicinity of the star. Since the gravitational field varies with elevation, these measurements will surely depend on r. The spherical symmetry guarantees that the measurements will be the same on all sides of the star, and thus they are independent of θ and ϕ. Let us also consider that we have stable conditions:

that the field is static during our observations, so that the measurements do not depend on t.

The metric is then not flat, but the 00 time–time component and the 11 space–space component must be modified by some functions of r. Thus it is of the form

$$ds^2 = B(r)c^2 dt^2 - A(r)dr^2 - r^2 d\theta^2 - r^2 \sin^2 \theta d\phi^2 \ , \qquad (3.1)$$

where $B(r)$ and $A(r)$ have to be found by solving the Einstein equations.

Far away from the star the spacetime is flat. This gives us the asymptotic conditions

$$\lim_{r \to \infty} A(r) = \lim_{r \to \infty} B(r) = 1 \ . \qquad (3.2)$$

From Eq. (2.77) the Newtonian limit of g_{00} is known. Here $B(r)$ plays the role of g_{00}; thus we have

$$B(r) = 1 - \frac{2GM}{c^2 r} \ . \qquad (3.3)$$

To obtain $A(r)$ from the Einstein equations is more difficult, and we shall not take the trouble of deriving it. The exact solution found by Karl Schwarzschild (1873–1916) in 1916 preceded any solution found by Einstein himself. The result is simply

$$A(r) = B(r)^{-1} \ . \qquad (3.4)$$

These functions clearly satisfy the asymptotic conditions (3.2).

Let us introduce the concept of the *Schwarzschild radius* r_c for a star of mass M, defined by $B(r_c) = 0$. It follows that

$$r_c \equiv \frac{2GM}{c^2} \ . \qquad (3.5)$$

The physical meaning of r_c is the following. Consider a body of mass m and radial velocity v attempting to escape from the gravitational field of the star. To succeed, the kinetic energy must overcome the gravitational potential. In the non-relativistic case the condition for this is

$$\tfrac{1}{2}mv^2 \geq GMm/r \ . \qquad (3.6)$$

The larger the ratio M/r of the star, the higher is the second cosmic velocity required to escape. Ultimately, in the ultra-relativistic case when $v = c$, only light can escape. At that point a non-relativistic treatment is no longer justified. Nevertheless, it just happens that the equality in (3.6) fixes the radius of the star correctly to be precisely r_c, as defined above. Thus nothing can escape from a star of smaller radius, not even light. Note that the escape velocity or second cosmic velocity of objects on Earth is 11 km/s, on the Sun it is 2.2×10^6 km/h, but on a black hole it is c.

This is the simplest kind of a *Schwarzschild black hole*, and r_c defines its *event horizon*. Inserting r_c in the functions A and B, and setting $d\theta = d\phi = 0$ for a spherically symmetric star, the *Schwarzschild metric* becomes

$$d\tau^2 = \left(1 - \frac{r_c}{r}\right) dt^2 - \left(1 - \frac{r_c}{r}\right)^{-1} \frac{dr^2}{c^2} \ . \tag{3.7}$$

The Schwarzschild metric has very fascinating consequences. Consider a space craft approaching a black hole with apparent velocity $v = dr/dt$ in the fixed frame of an outside observer. Light signals from the space craft travel on the light cone, $d\tau = 0$, so that

$$\frac{dr}{dt} = c\left(1 - \frac{r_c}{r}\right) \ . \tag{3.8}$$

Thus the space craft appears to slow down with decreasing r, finally coming to a full stop as it reaches $r = r_c$.

No information can ever reach the outside observer beyond the event horizon. The reason for this is the mathematical singularity of dt in the expression

$$c\,dt = \frac{dr}{1 - r_c/r} \ . \tag{3.9}$$

The time intervals dt between successive crests in the wave of the emitted light become longer, reaching infinite wavelength at the singularity. Thus the frequency ν of the emitted photons goes to zero, and the energy $E = h\nu$ of the signal vanishes. One cannot receive signals from beyond the event horizon because photons cannot have negative energy. Thus the outside observer sees the space craft slowing down and the signals redshifting until they cease completely.

The pilot in the space craft uses local coordinates, so he sees the passage into the black hole entirely differently. If he started out at distance r_0 with velocity $dr/dt = 0$ at time t_0, he will have reached position r at proper time τ which we can find by integrating $d\tau$ in Eq. (3.7) from 0 to τ,

$$\int_0^\tau \sqrt{d\tau^2} = \tau = \int_{r_0}^r \left[\frac{1 - r_c/r}{(dr/dt)^2} - \frac{1}{c^2(1 - r_c/r)}\right]^{\frac{1}{2}} dr \ . \tag{3.10}$$

The result depends on $dr(t)/dt$ which can only be obtained from the equation of motion. The pilot considers that he can use Newtonian mechanics, so he may take

$$\frac{dr}{dt} = c\sqrt{\frac{r_c}{r}} \ .$$

The result is then

$$\tau \propto (r_0 - r)^{3/2}. \tag{3.11}$$

However, many other expressions for dr/dt also make the integral in Eq. (3.10) converge.

Thus the singularity at r_c does not exist to the pilot, his comoving clock shows finite time when he reaches the event horizon. Once across it he reaches the centre of the black hole rapidly. For a black hole of mass $10\,M_\odot$ this final passage lasts about 10^{-4} s.

Although this space craft voyage is pure science fiction, we may be able to observe the collapse of a supernova into a black hole. It seems now settled that a sufficiently massive star, at the endpoint of its stellar evolution, will undergo gravitational collapse to form a black hole. Just as for the space craft, the collapse towards the Schwarzschild radius will appear to take a very long time. Towards the end of it the ever redshifting light will fade and finally disappear completely.

Note from the metric equation (3.7) that inside r_c the time term becomes negative and the space term positive, thus space becomes timelike and time spacelike. The implications of this are best understood if one considers the shape of the light cone of the space craft during its voyage, see Fig. 14. Outside the event horizon the future light cone contains the outside observer who receives signals from the space craft. Nearer r_c the light cone narrows and the slope dt/dr steepens because of the approaching singularity in the righthand side expression of Eq. (3.9). The portion of the future spacetime which can receive signals therefore diminishes.

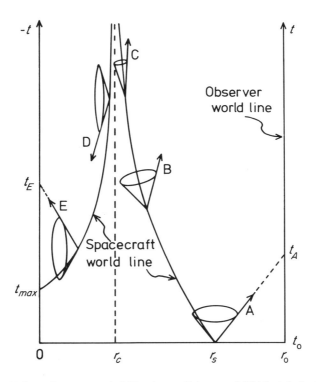

Fig. 14 The world line of a spacecraft falling into a Schwarzschild black hole. (A): The journey starts at time t_0 when the spacecraft is at a radius r_s, far outside the Schwarzschild radius r_c, and the observer is at r_0. A light signal from the spacecraft reaches the observer at time $t_A > t_0$ (read time on the righthand vertical scale!). (B): Nearer the black hole the future light cone of the spacecraft tilts inward. A light signal along the arrow will still reach the observer at a time $t_B \gg t_A$. (C): Near the Schwarzschild radius the light cone narrows considerably, and a light signal along the arrow reaches the observer only in a very distant future. (D): Inside r_c the time and space directions are interchanged, time running from up to down on the lefthand vertical scale. All light signals will reach the centre of the black hole at $r=0$, and none will reach the observer. The arrow points in the backward direction, so that a light signal will reach the centre after the spacecraft. (E): The arrow points in the forward direction of the black hole, so that a light signal will reach the centre at time t_E, which is earlier than t_{max} when the spacecraft ends its journey

After crossing the event horizon the future light cone is turned inwards because the time and space axes have exchanged positions. Thus no part of the outside spacetime is included in the cone. The slope of the light cone is vertical at the horizon. Thus it defines at the same time a cone of zero opening angle around the original time axis, and a cone of 180° around the final time axis, encompassing the full spacetime of the black hole. As the space craft approaches the centre, dt/dr decreases, defining a narrowing opening angle which always contains the centre. When the centre is reached the space craft has no future anymore.

In quantum mechanics it is always possible to associate the mass of a particle M with a wave having the *Compton wavelength*

$$\lambda = \frac{\hbar}{Mc} .$$ (3.12)

In other words, for a particle of mass M quantum effects become important at distances of the order of λ. On the other hand, gravitational effects are important at distances of the order of the Schwarzschild radius. Equating the two distances we find the scale at which quantum effects and gravitational effect are of equal importance. This defines the *Planck mass*

$$M_P = \sqrt{\hbar c/G} = 1.221 \times 10^{19} \text{ GeV}/c^2 .$$ (3.13)

From this we can derive the Planck energy $M_P c^2$ and the *Planck time*

$$t_P = \lambda_P/c = 5.31 \times 10^{-44} \text{ s} .$$ (3.14)

We shall later on make frequent use of quantities at the Planck scale. The reason for associating these scales with Planck's name is that he was the first to notice that the combination of fundamental constants

$$\lambda_P = \sqrt{\hbar G/c^3} = 1.62 \times 10^{-35} \text{ m}$$ (3.15)

yielded a natural scale of length.

Black holes may have been created in the Big Bang, and they are created naturally in the aging of stars. Red giants have a centre resisting further compactification because there is a limit to how densely electrons can be packed; this is called the electron *degeneracy pressure*. If the mass of this centre exceeds the *Chandrasekhar limit* of $1.4 M_\odot$, its mass density is so high that the electron degeneracy pressure cannot withstand the huge force of gravity. When the star has burned all its fuel it will then collapse to a neutron star, stabilized by the neutron degeneracy pressure. However, if the mass exceeds the *Landau–Oppenheimer–Volkov limit* of $2.3 M_\odot$, even the neutron star collapses further to become a black hole.

Stephen Hawking has shown [1] that although no light can escape from black holes, they can nevertheless radiate if one takes quantum mechanics into account. It is a property of the *vacuum* that particle–antiparticle pairs such as $e^- e^+$ are continuously created out of nothing, to disappear in the next moment by *annihilation* which is the inverse process. Since energy cannot be created nor destroyed, one of the particles must have positive energy and the other one an equal amount of negative energy. They form a *virtual pair*, neither one is real in the sense that it could escape to infinity or be observed by us.

In a strong electromagnetic field the electron e^- and the positron e^+ may become separated by a Compton wavelength λ of the order of the Schwarzschild radius r_c. Hawking has shown that there is a small but finite probability for one of them to 'tunnel' through the barrier of the quantum vacuum and escape the black hole horizon as a real particle with positive energy, leaving the negative energy particle inside the horizon of the black hole. Since energy must be conserved the black hole loses mass in this process, called *Hawking radiation*. The rate of particle emission is as if the black hole were a hot body of temperature proportional to the surface gravity. The timescale of complete evaporation is

$$t \approx 10 \text{ Gyr} \left(\frac{M}{10^{12} \text{ kg}} \right)^3 . \tag{3.16}$$

Thus small black holes evaporate fast, whereas heavy ones may have lifetimes exceeding the age of the Universe.

Black holes are very simple objects as seen from outside their event horizon, they have only the three properties: mass, electric charge and angular momentum. Their size depends only on their mass so that all black holes with the same mass are identical and exactly spherical, unless they rotate. All other properties possessed by stars, such as shape, electric dipole moment, magnetic moments as well as any detailed outward structure are absent. This has led to the famous statement 'black holes have no hair'.

So far we have studied only the simplest kind of *Schwarzschild black holes* which have mass but no electric charge, and which do not rotate. Black holes possessing either charge or angular momentum are called *Reissner–Nordström black holes* and *Kerr black holes*, respectively, and they are described by different metrics. The additional energy possessed by rotating black holes leads to a smaller event horizon. If the angular momentum of the star is sufficiently large it may even overcompensate the gravitational binding energy. Then there is no event horizon and we have the case of a visible singularity, also called a *naked singularity*. The physics of such an object is not understood, but the reader might find further enjoyment reading the popular book by Hawking on this subject [2].

The inside of black holes are subjects of pure speculation because of the difficulties of solving the Einstein equations. In the Schwarzschild type there is of course the singularity at the centre. In the other types there seem to be ways to avoid the singularity, so that matter or radiation falling in might 'tunnel' through a 'wormhole' out into another universe. Needless to say, all such ideas are purely theoretical with no hope of experimental verification.

At the time of writing (1993) astronomers believe they have identified a few objects harbouring a black hole. In each of these the evidence is a strong X-ray signal from the vicinity of a binary star. The enormous gravitational pull of the black hole tears material from its companion star. This material then orbits the black hole in an Saturnus-like accretion disc before disappearing into the black hole. Gravity and friction heat the material in the accretion disc until it emits X-rays. Present candidates comprise the LMC X-1 in the Large Magellanic Cloud, the Cygnus X-1, the Cygnus V404, the A0620-00 in the constellation Monoceros, and perhaps the Nova Muscae 1991.

3.2 Friedman Cosmologies

The case of a spherically symmetric body treated in the previous section is, however, not a good model for our expanding Universe. Let us therefore return to the homogeneous and isotropic universe for which the Robertson–Walker metric (2.35) was derived. Recall that it had only diagonal components, Eq. (2.52), and that it contained the curvature parameter k.

The stress-energy tensor $T_{\mu\nu}$ entering on the right-hand side of Einstein's equations (2.82) was given by Eq. (2.78) in its diagonal form. For a comoving observer with velocity four-vector $v = (c,0,0,0)$ the time–time component T_{00} and the space–space component T_{11} are then

$$T_{00} = \rho c^2 , \quad T_{11} = \frac{pS^2}{1 - k\sigma^2} , \tag{3.17}$$

taking g_{11} from Eq. (2.52). We will not need any other components of $T_{\mu\nu}$. In what follows we shall denote mass density ρ and energy density ρc^2 or ε.

On the left-hand side of Einstein's equations (2.82) we need G_{00} and G_{11} to equate with T_{00} and T_{11}, respectively. We have all the tools to do it: the g_{00} and g_{11} given in Eq. (2.52) are inserted into the expressions for the Ricci tensor $R_{\mu\nu}$, Eq. (2.63), and the Ricci scalar R, Eq. (2.64). The result is

$$G_{00} = 3(cS)^{-2}(\dot{S}^2 + kc^2) \tag{3.18}$$

$$G_{11} = -c^{-2}(2S\ddot{S} + \dot{S}^2 + k)(1 - k\sigma^2)^{-1} . \tag{3.19}$$

Substituting Eqs. (3.17–3.19) into Einstein's equations (2.82) we obtain two distinct dynamical relations for the cosmic scale factor $S(t)$,

$$\frac{\dot{S}^2 + kc^2}{S^2} = \frac{8\pi}{3} G\rho , \tag{3.20}$$

$$\frac{2\ddot{S}}{S} + \frac{\dot{S}^2 + kc^2}{S^2} = -\frac{8\pi}{c^2} Gp . \tag{3.21}$$

These equations were derived in 1922 by the Russian physicist and mathematician *Alexandr Friedman* (1888–1925) seven years before Hubble's discovery, at a time when even Einstein did not believe in his own equations because they did not allow the Universe to be static. As we shall see, the expansion (or contraction) of the Universe is inherent to Friedman's equations.

Equation (3.20) shows that the rate of expansion, $\dot{S}$, increases with the mass density ρ in the Universe. Subtracting it from the second equation we obtain

$$\frac{2\ddot{S}}{S} = -\frac{8\pi}{3c^2} G(\rho c^2 + 3p) . \tag{3.22}$$

This shows that the acceleration of the expansion decreases with increasing pressure and energy density, whether mass or radiation energy. Thus it is more appropriate to talk about the *deceleration* of the expansion.

At our present time t_0 when the mass density is ρ_0, the cosmic scale is S_0, the Hubble parameter is H_0 and the density parameter Ω_0 is given by Eq. (2.10), Friedman's equation (3.20) takes the form

$$\dot{S}_0^2 = \frac{8\pi}{3} G S_0^2 \rho_0 - kc^2 = H_0^2 S_0^2 \Omega_0 - kc^2 \ . \tag{3.23}$$

It is interesting to note that this reduces to the Newtonian relation (2.12) if we make the identification

$$kc^2 = H_0^2 S_0^2 \ (\Omega_0 - 1) \ . \tag{3.24}$$

Thus the relation between the Robertson–Walker curvature parameter k and the present density parameter Ω_0 emerges: to the k values +1, 0, and −1 correspond an overcritical density $\Omega_0 > 1$, a critical density $\Omega_0 = 1$ and an undercritical density $0 < \Omega_0 < 1$, respectively. The relation (3.24) also follows if we substitute $\dot{S} = HS$ from Eq. (2.45) into Eq. (3.20).

The spatial curvature R introduced in Eq. (2.64) can be expressed in terms of Ω,

$$R = \frac{6k}{S^2} = 6H^2(\Omega - 1) \ . \tag{3.25}$$

Obviously R vanishes in a flat universe, and it is only meaningful when it is non-negative as in a closed universe. It is conventional to define a 'radius of curvature' valid also for open universes, by

$$r_U \equiv \frac{S}{\sqrt{|k|}} = \sqrt{\frac{6}{R}} = \frac{1}{H\sqrt{|\Omega - 1|}} \ . \tag{3.26}$$

For a closed universe r_U has the physical meaning of the radius of a sphere.

Friedman's equations did not gain general recognition until after Friedman's death, when they were confirmed by an independent derivation in 1927 by Georges Lemaître (1894–1966). For now they will constitute the tools for our further investigations. Let us first consider the static universe cherished by Einstein. This implies that $S(t)$ is a constant S_0 so that $\dot{S} = 0$ and $\ddot{S} = 0$, and the age of the Universe is infinite. The equations (3.20) and (3.21) then reduce to

$$\frac{kc^2}{S_0^2} = \frac{8\pi}{3} G \rho_0 = -\frac{8\pi}{c^2} G p_0 \ . \tag{3.27}$$

In order that the mass density ρ_0 today be positive, k must be +1. Note that this leads to the surprising result that the pressure of matter p_0 becomes negative!

Einstein corrected for this in 1917 by introducing a constant Lorentz-invariant term $\lambda g_{\mu\nu}$ into his law of gravitation, Eq. (2.82). In contrast to the two terms in Eq. (2.81) making up the Einstein tensor $G_{\mu\nu}$, the $\lambda g_{\mu\nu}$ term does not vanish in the limit of flat spacetime. This *cosmological constant* λ corresponds to a tiny but universal force acting on matter. With this addition Friedman's equations take the form

$$\frac{\dot{S}^2 + kc^2}{S^2} - \frac{\lambda}{3} = \frac{8\pi}{3}G\rho \, , \tag{3.28}$$

$$\frac{2\ddot{S}}{S} + \frac{\dot{S}^2 + kc^2}{S^2} - \lambda = -\frac{8\pi}{c^2}Gp \, . \tag{3.29}$$

The cosmology described by these equations is called the *Friedman–Lemaître–Robertson–Walker Universe*. A positive value of λ corresponds to a repulsive force counteracting the conventional attractive gravitation, as required by Einstein. The case when λ is adjusted to give a static solution is called the *Einstein Universe*. A negative λ corresponds to an additional attractive force.

The pressure of matter is certainly very small, otherwise one would observe the galaxies having random motion similar to that of molecules in a gas under pressure. Thus one can set $p = 0$ to a good approximation. In the static case when $S = S_0$, $\dot{S} = 0$ and $\ddot{S} = 0$, Eq. (3.28) becomes

$$\frac{kc^2}{S_0^2} - \frac{\lambda}{3} = \frac{8\pi}{3}G\rho_0 \, .$$

It follows from this that in a flat Universe where $k = 0$,

$$\frac{\lambda}{8\pi G} = -\rho_0 \, . \tag{3.30}$$

The quantity on the left is called the *vacuum energy density*. Using this relation we can find the value of λ corresponding to the attractive force of the present mass density,

$$-\lambda = 8\pi G\rho_0 = 3\Omega_0 H_0^2 \gtrsim \frac{1}{3}10^{-34}\text{s}^{-2} \approx 4 \times 10^{-52}c^2 \text{ m}^{-2} \tag{3.31}$$

Thus we can infer from the expansion that $|\lambda|$ must be smaller than this.

No quantity in physics this small has ever been known before. It is comforting that the universal repulsive force acting on matter is so small, but it is extremely uncomfortable that λ differs from zero only in the 52nd decimal (in units of $c = 1$). It would be much more natural if λ were exactly zero. This situation is one of the enigmas which will remain with us to the end of this book. As we shall see, the repulsive force may have been of great importance during the first brief moments of the existence of the Universe.

When Hubble discovered the expansion in 1929 Einstein abandoned his belief in a static universe, withdrawing the cosmological constant and calling it 'the greatest blunder of my lifetime'. Later development has reintroduced λ into a non-static cosmology, thus accepting Einstein's conjecture, but for an entirely different reason. The currently most popular cosmological model assumes no cosmological constant, a negligible pressure and flat space. This is called the *Einstein–de Sitter Universe*.

Let us return to study the solutions of Friedman's equations with $\lambda = 0$ in the general case of non-vanishing pressure p. We assume that p is varying slowly enough for $\dot{p}$ to be neglected in comparison with ρ and $\dot{S}$. Differentiating Eq. (3.20) with respect to time,

$$\frac{d}{dt}(\dot{S}^2 + kc^2) = \frac{8\pi}{3}G\frac{d}{dt}(\rho S^2) \, ,$$

we obtain an equation of second order in the time derivative,

$$2\dot{S}\ddot{S} = \frac{8\pi}{3}G(\dot{\rho}S^2 + 2\rho S\dot{S}) \, . \tag{3.32}$$

Using Eq. (3.22) to cancel the second-order time derivative and multiplying through by S, we obtain a new equation containing only first-order time derivatives,

$$\dot{\rho}c^2 S^3 + 3(\rho c^2 + p)S^2\dot{S} = 0 \, . \tag{3.33}$$

This equation can easily be integrated,

$$\int \frac{\dot{\rho}(t)c^2}{\rho(t)c^2 + p(t)}dt = -3 \int \frac{\dot{S}(t)}{S(t)}dt \tag{3.34}$$

if we know the relation between energy density and pressure, the *equation of state* of the Universe.

Note that all terms in Eq. (3.33) now have dimension energy/time. In other words, this equation states that the change of energy per time is zero, so we can interpret it as the *law of conservation of energy*. As we have seen, it follows directly from Friedman's equations without any further assumptions.

In contrast, the *law of conservation of entropy* is not implied by Friedman's equations. If we assume that the expansion is *adiabatic* during the radiation-dominated epoch as well as during the matter-dominated epoch (we shall come back to this assumption in Section 4.2), then it follows that entropy s is conserved,

$$\dot{s} = 0 \, . \tag{3.35}$$

Then we can make an Ansatz for the equation of state: let p be proportional to ρc^2 with some proportionality factor α which is a constant in time,

$$p = \alpha\rho \, . \tag{3.36}$$

Here we have absorbed c^2 in the definition of α. In fact, one can show that this is the most general equation of state in a spacetime with the Robertson–Walker metric. Inserting this Ansatz into the integral in Eq. (3.34) we find that the relation between energy density and scale is

$$\rho \propto S^{-3(1+\alpha)} \, . \tag{3.37}$$

The value of the proportionality factor α follows from the adiabaticity condition. Leaving the derivation of α for a more detailed discussion in Section 4.2, we shall anticipate here its values in three special cases of great importance:

(i) A *matter-dominated* universe filled with non-relativistic cold matter for which it is a good approximation to set $p = 0$. From Eq. (3.36) then, this corresponds to $\alpha = 0$, and Eq. (3.37) becomes

$$\rho \propto S^{-3} . \tag{3.38}$$

(ii) A *radiation-dominated* universe filled with an ultra-relativistic hot gas composed of non-interacting particles of energy density ε. Then statistical mechanics tells us that the equation of state is

$$p = \frac{1}{3}\varepsilon = \frac{1}{3}\rho c^2 . \tag{3.39}$$

This evidently corresponds to $\alpha = \frac{1}{3}$, so that the density is

$$\rho \propto S^{-4} . \tag{3.40}$$

(iii) The *vacuum energy* state corresponds to a flat, static universe ($\ddot{S} = 0$, $\dot{S} = 0$, $k = 0$) with a cosmological term. From Eqs. (3.28) and (3.29) we then obtain

$$p = -\rho c^2 , \quad \alpha = -1 . \tag{3.41}$$

Thus the pressure of the vacuum energy is negative in agreement with the definition in Eq. (3.30) of the vacuum energy density as a negative quantity. The equation of state is then

$$\rho \propto \text{const.} \tag{3.42}$$

It follows from the S-dependence of ρ in the cases (i) and (ii) that the curvature term obeys the following inequality in the limit of small S:

$$\frac{kc^2}{S^2} \ll \frac{8\pi}{3}G\rho . \tag{3.43}$$

In fact, the inequality is always true when

$$p > -\frac{1}{3}\rho c^2 , \quad \alpha > -\frac{1}{3} . \tag{3.44}$$

Then we can neglect the curvature term in Eq. (3.20), which simplifies to

$$\frac{\dot{S}}{S} = H(t) = \left(\frac{8\pi}{3}G\rho\right)^{\frac{1}{2}} \propto S^{-3(1+\alpha)/2} . \tag{3.45}$$

Let us now find the time-dependence of S by integrating this differential equation:

$$\int dS S^{-1+3(1+\alpha)/2} \propto \int dt .$$

Supposing that the integrals can be carried out from the lower limit $S = 0$ at a time t_{start}, the solution is

$$\frac{2}{3(1 + \alpha)} S^{3(1+\alpha)/2} \propto t - t_{start} .$$

Choosing $t_{start} = 0$, we have

$$S(t) \propto t^{2/3(1+\alpha)} . \tag{3.46}$$

In the two special cases of matter-domination and radiation-domination we know the value of α. Inserting this into Eq. (3.46) we obtain the time-dependence of S for the present matter-dominated universe,

$$S(t) \propto t^{2/3} , \tag{3.47}$$

and for the radiation-dominated universe at very early times,

$$S(t) \propto t^{1/2} . \tag{3.48}$$

Independently of the value of k in the curvature term neglected above we find the starting value of the scale,

$$\lim_{t \to 0} S(t) = 0 . \tag{3.49}$$

Actually we do not even need an equation of state to arrive at this limit. Provided $\rho c^2 + 3p$ was always positive, as it is today, we can see from Eq. (3.22) that the Universe has always decelerated. It follows then that S must have been zero at some time in the past. This time we have already chosen as $t = 0$.

In the same limit the rate of scale change $\dot{S}$ is obtained from Eq. (3.45) with $\alpha = \frac{1}{3}$,

$$\lim_{t \to 0} \dot{S}(t) = \lim_{t \to 0} S^{-1}(t) = \infty . \tag{3.50}$$

It follows from Eqs. (3.39) and (3.40) that early times were characterized by extreme density and pressure,

$$\lim_{t \to 0} \rho(t) = \lim_{t \to 0} S^{-4}(t) = \infty ,$$

$$\lim_{t \to 0} p(t) = \lim_{t \to 0} S^{-4}(t) = \infty .$$

The time $t = 0$ is therefore called the *Big Bang*, as first proposed by Lemaître in 1932. Whether Friedman's equations in fact can be trusted at that limit is another story which we shall come back to later.

Note that the inequality (3.43) is always true in a flat universe, $k = 0$, or open one, $k = -1$. Then the expansion always continues, following Eq. (3.46). In a closed universe, $k = +1$, the density drops with the third power of S according to Eq. (3.38), whereas the curvature term kc^2/S^2 only drops with the second power, so the inequality (3.43) is finally violated. This happens at a scale S_{max} such that

$$S_{max}^{-2} = \frac{8\pi G\rho}{3c^2} , \tag{3.51}$$

and the expansion halts because $\dot{S} = 0$ in Eq. (3.20). Let us call this the turnover time t_{max}. At later times the expansion turns into contraction, and the Universe returns to zero size at time $2t_{max}$. That time may be called the *Big Crunch*.

The three cases $k = -1, 0, +1$ are illustrated qualitatively in Fig. 15. Note that we have to require that all models are consistent with the scale and rate of expansion today, S_0 and $\dot{S}_0$ at time t_0. Following the curves back in time one notices that they intersect the time axis at different times. Thus what may be called time $t = 0$ is more recent in a flat universe than in an open universe, and in a closed universe it is even more recent.

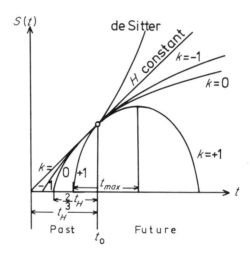

Fig. 15 Time dependence of the cosmic scale $S(t)$ in various scenarios, all of which correspond to the same constant slope $H=H_0$ at the present time t_0. $k=+1$: A closed universe with a total lifetime $2t_{max}$. It started more recently than would have a flat universe. $k=0$: A flat universe which started $\frac{2}{3}t_H$ ago. $k=-1$: An open universe which started at a time $\frac{2}{3}t_H < t < t_H$ before the present time. de Sitter: An exponential (inflationary) scenario corresponding to a large cosmological constant. This is also called the Lemaître cosmology

Let us calculate the values of S_{max} and t_{max} in a closed universe. From Friedman's equation (3.20) with $k = +1$ we have

$$\frac{dS}{dt} = \sqrt{\frac{8\pi}{3}G\rho(S)S^2 - c^2} .$$

Thus t_{max} is obtained by integrating t from 0 to t_{max} and S from 0 to S_{max},

$$t_{max} = \frac{1}{c}\int_0^{S_{max}} dS \left(\frac{8\pi G}{3c^2}\rho(S)S^2 - 1\right)^{-\frac{1}{2}} . \tag{3.52}$$

To solve the S-integral we need to know the energy density $\rho(S)$ in terms of the scale factor, and we need to know S_{max}. Let us take the mass of the Universe to be M. We have already found in Eq. (2.41) that the volume of a closed universe with Robertson–Walker metric is

$$V = 2\pi^2 S^3 \ .$$

In a matter-dominated universe the energy density is mostly mass density, thus

$$\rho = \frac{M}{V} = \frac{M}{2\pi^2 S^3} \ . \tag{3.53}$$

This agrees perfectly with the result (3.38) that the density in a matter-dominated universe is inversely proportional to S^3. Obviously the missing proportionality factor in Eq. (3.38) is then $M/2\pi^2$. Inserting the density (3.53) with $S = S_{max}$ into Eq. (3.51) we obtain

$$S_{max} = \frac{4MG}{3\pi c^2} \ . \tag{3.54}$$

We can now complete the integral in Eq. (3.52) with ρ from Eq. (3.53) and S_{max} from (3.54), and solve it:

$$t_{max} = \frac{\pi}{2c} S_{max} = \frac{2MG}{3c^3} \tag{3.55}$$

Although we do not know whether we live in a closed universe we certainly know from the ongoing expansion that $t_{max} > t_0$. Using the lower limit for t_0 from Eq. (1.23) we find a lower limit to the mass of the Universe,

$$M > \frac{3t_0 c^3}{2G} \approx 10^{23} \ M_\odot \ . \tag{3.56}$$

Actually the total mass inside the present horizon is estimated to be about $10^{22} \ M_\odot$.

The dependence of t_{max} on Ω can easily be obtained from Eq. (3.51),

$$t_{max} = \frac{\pi}{2c} \left(\frac{8\pi G}{3c^2} \rho_c \Omega \right)^{-\frac{1}{2}} = \frac{\pi}{2H_0 \sqrt{\Omega}} \ . \tag{3.57}$$

Thus we see that in the limiting case for a closed universe when $\Omega = 1$ the value of t_{max} is $\pi/2$ times the Hubble time.

Another interesting quantity is the Schwarzschild radius of the Universe. Combining Eqs. (3.55) and (3.5) we find

$$r_{c,\text{Universe}} = 3ct_{max} > 12 \text{ Gpc.} \tag{3.58}$$

Comparing this number with the much smaller Hubble radius $3h^{-1}$ Gpc in Eq. (1.24) we might conclude that we live inside a black hole! However, the Schwarzschild metric is static whereas the Hubble radius recedes in expanding Friedman models at a velocity

$$\dot{r}_H = c(1 + q) > c \, , \tag{3.59}$$

so it will catch up with $r_{c,\text{Universe}}$ at some time.

Above we only integrated Eq. (3.20) in order to obtain the turnover time of a closed matter-dominated universe. In greater generality, we could ask for the time t when a closed, flat or open universe has reached size $S = S_0(1 + z)$ under the assumptions of matter-domination or radiation-domination. In each case the integral can be solved analytically, but we shall not quote all the possible cases (see reference [3]).

In a flat matter-dominated universe (with $k = 0$, $\Omega = 1$) the time t_z corresponding to a redshift z is

$$t_z = \frac{2}{3H_0}(1 + z)^{-3/2} \, . \tag{3.60}$$

Thus the present age of the Universe at $z = 0$ would be

$$t_0 = \frac{2}{3H_0} \, , \tag{3.61}$$

as shown in Fig. 15 for the curve of a flat universe. In that case the size of the Universe would be $ct_0 = 2h^{-1}$ Gpc.

In an open matter-dominated universe (with $k = -1$, $\Omega < 1$) the age of the Universe is given by the expression

$$t_0 = \frac{1}{2H_0} \left[\frac{2}{1 - \Omega_0} - \frac{\Omega_0}{(1 - \Omega_0)^{3/2}} \cosh^{-1}(2\Omega_0^{-1} - 1) \right] \, . \tag{3.62}$$

This is plotted in Fig. 16 for several values of the parameter h. Note that the only time scale available is the inverse of the Hubble parameter. This is a consequence of the fact that Friedman's equations are invariant under time translations.

The limit $\Omega = 0$ represents the slowest possible expansion. This permits us to set an upper limit to the present age of the Universe,

$$t_0 < H_0^{-1} \approx 10h^{-1} \text{ Gyr} \, . \tag{3.63}$$

Near $\Omega = 1$ the expression (3.62) can be expanded into the series

$$t_0 = \frac{2}{3H_0}[1 - \tfrac{1}{5}(\Omega - 1) + ...] \, . \tag{3.64}$$

It follows from Eqs. (3.62) and (3.64) that t_0 is a decreasing function of Ω: the more matter there is in the Universe, the more important is the deceleration, and the younger is the Universe. Thus a known lower limit to t_0 like the one quoted in Eq. (1.23) can be used to set an upper limit to Ω_0. It is then necessary to use a lower limit also to h. Using $h = 0.5$ from Table 2 on page 6 we obtain

$$\Omega_0 \leq 1.5 \, . \tag{3.65}$$

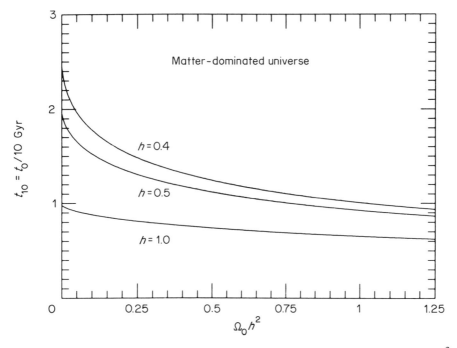

Fig. 16 The age of a matter-dominated universe as a function of the density parameter Ωh^2 for different values of h, from reference [3] by permission of E.W. Kolb and M. Turner. One nearly always needs to account for h when a cosmological parameter has to be specified numerically

A closed universe is always decelerating. We do not know how large the rate of expansion has been initially, but we can deduce from Eqs. (3.20) and (3.24) that as t decreases $\dot{S}$ increases, and the dimensionless quantity

$$\frac{kc^2}{\dot{S}^2} = \Omega - 1 \tag{3.66}$$

decreases. Thus it appears that Ω *may have been arbitrarily close to* 1 at some early time regardless of its present value.

3.3 de Sitter Cosmologies

Let us now turn to another special case for which Einstein's equations can be solved. Consider a homogeneous *flat* universe with the Robertson–Walker metric in which the density of matter is a constant, $\rho(t) = \rho_0$. Friedman's equation (3.28) for the rate of expansion including the cosmological constant then takes the form

$$\frac{\dot{S}(t)}{S(t)} = H , \tag{3.67}$$

where H is now a constant,

$$H = \sqrt{\frac{8\pi}{3}G\rho_0 + \frac{\lambda}{3}} \;.$$

(3.68)

The solution to Eq. (3.67) is obviously an exponentially expanding Universe,

$$S(t) \propto e^{Ht} \;.$$

(3.69)

This is drawn as the de Sitter curve in Fig. 15. Substituting this function into the Robertson–Walker metric (2.35) we obtain

$$ds^2 = c^2 dt^2 - e^{2Ht}\left(d\sigma^2 + \sigma^2 d\theta^2 + \sigma^2 \sin^2\theta d\phi^2\right) \;.$$

(3.70)

In 1917 *Willem de Sitter* (1872–1934) published such a solution, setting $\rho = p = 0$, thus relating H directly to the cosmological constant λ. The same solution of course follows even without λ if the density ρ is constant. Thus $\lambda/8\pi G$ represents an energy density, as we have already noted in Eq. (3.30). In flat spacetime the de Sitter metric is given by Eq. (3.70).

Eddington characterized the *de Sitter universe* as 'motion without matter' in contrast to the static *Einstein universe* which was 'matter without motion'. If one introduces two test particles into this empty de Sitter universe they will appear to recede from each other exponentially. The force driving the test particles apart is very strange. Let us suppose that they are at spatial distance rS from each other, and that λ is positive. Then the equation of relative motion of the test particles is given by Eq. (3.22) including the λ term,

$$\frac{d^2(rS)}{dt^2} = \frac{\lambda}{3}rS - \frac{4\pi}{3}G(\rho + 3pc^{-2})RS \;.$$

(3.71)

The second term on the right is the decelerating force due to the ordinary gravitational interaction. The first term, however, is a force due to the vacuum energy density, proportional to the distance r between the particles!

If λ is positive as in the Einstein universe, the force is repulsive, accelerating the expansion. If λ is negative the force is attractive, decelerating the expansion just like ordinary gravitation. This is called an *anti-de Sitter* universe. Since λ is so small (*cf.* Eq. (3.31)) this force will only be of importance to systems with mass densities of the order of the vacuum energy. The only known systems with such low densities are the large-scale structures, or the full horizon volume of cosmic size. This is the reason for the name *cosmological constant*.

Universes with exponential expansion are nowadays called *inflationary*. Although we started by considering a flat universe, inflation is not restricted to k=0 spacetimes. For a universe with $k = +1$ one finds the solution

$$S(t) \propto H^{-1}\cosh Ht \;,$$

(3.72)

and for a universe with negative curvature, $k = -1$, one finds

$$S(t) \propto H^{-1}\sinh Ht \;.$$

(3.73)

The metric is as in the flat case with the above expressions replacing exp Ht in Eq. (3.70). In both these cases $S(t)$ obviously tends to the exponential solution (3.69) for sufficiently late times, $Ht \gg 1$. In contrast, a Friedman universe obeys $H_0 t_0 < 1$ today, according to the condition (3.63).

As we already could see from Eq. (3.68), the Hubble parameter in flat space is constant,

$$H = \sqrt{\frac{\lambda}{3}} \ . \tag{3.74}$$

In curved space with $k = \pm 1$, however, it is a function of time

$$H = \sqrt{\frac{\lambda}{3}} \ \tanh^k \left(\sqrt{\frac{\lambda}{3}} t \right) \ . \tag{3.75}$$

Actually, the cases $k = -1, 0, +1$ correspond to different branches of a five-dimensional Minkowski spacetime with the coordinates

$$
\begin{aligned}
z_0 &= H^{-1} \sinh Ht + \tfrac{1}{2} H e^{Ht} \sigma^2 \ , \\
z_1 &= H^{-1} \cosh Ht - \tfrac{1}{2} H e^{Ht} \sigma^2 \ , \\
z_2 &= e^{Ht} \sigma \sin \theta \cos \phi \ , \\
z_3 &= e^{Ht} \sigma \sin \theta \sin \phi \ , \\
z_4 &= e^{Ht} \sigma \cos \theta \ .
\end{aligned}
\tag{3.76}
$$

The de Sitter space may then be represented by the hyperboloid

$$z_0^2 - z_1^2 - z_2^2 - z_3^2 - z_4^2 = -H^{-2}. \tag{3.77}$$

The branch corresponding to flat Robertson–Walker spacetime spans the half-hyperboloid defined by $z_0 + z_1 > 0$.

Although the world is not devoid of matter, the de Sitter universe may still be of more than academical interest in situations when ρ changes much slower than the scale S. Let us replace σ by r in Eqs. (3.76) and redefine z_0, z_1 as follows:

$$
\begin{aligned}
z_0 &= \sqrt{H^{-2} - r^2} \sinh Ht \ , \\
z_1 &= \sqrt{H^{-2} - r^2} \cosh Ht \ .
\end{aligned}
\tag{3.78}
$$

The de Sitter metric then takes the form

$$ds^2 = (1 - r^2 H^2) dt^2 - (1 - r^2 H^2)^{-1} dr^2 - r^2 (d\theta^2 + \sin^2\theta d\phi^2) \ , \tag{3.79}$$

which resembles the Schwarzschild metric, Eq. (3.7). There is an *inside* region in the de Sitter space at $r < H^{-1}$, for which the metric tensor component g_{00} is positive and g_{11} is negative. This resembles the region *outside* a black hole of Schwarzschild radius $r_c = H^{-1}$, at $r > r_c$ where g_{00} is positive and g_{11} is negative. Outside the radius $r = H^{-1}$ in de Sitter space and inside the Schwarzschild black hole these components of the metric tensor change sign.

The interpretation of this geometry is that the de Sitter metric describes an expanding spacetime surrounded by a black hole. Inside the region $r = H^{-1}$ no signal can be received from distances outside H^{-1} because there the metric corresponds to the inside of a black hole! In an anti-de Sitter universe the constant attraction ultimately dominates, so that the expansion turns into contraction. Thus de Sitter universes are open and anti-de Sitter universes are closed.

Let us study the particle horizon r_H in a de Sitter universe. Recall that this is defined as the location of the most distant visible object, and that the light from it started on its journey towards us at time t_H. From Eqs. (2.46) and (2.48) the particle horizon is at

$$r_H = S(t) \int_{t_H}^{t_0} \frac{dt'}{PS(t')} \ . \tag{3.80}$$

The scale factor $S(t)$ in the spatially flat de Sitter universe is $\exp Ht$. Let us choose t_H as the origin of time, $t_H = 0$. The distance $r_H(t)$ as a function of the time of observation t then becomes

$$r_H(t) = H^{-1} e^{Ht}(1 - e^{-Ht}) \ . \tag{3.81}$$

The comoving distance to the particle horizon, $\sigma_p = r_H/S$, approaches quickly the constant value H^{-1}. Thus for a comoving observer in this world the particle horizon would *always* be located at H^{-1}. Points which were inside this horizon at some time will be able to exchange signals, but events outside the horizon cannot influence anything inside this world.

The situation in a Friedman universe is quite different. There the time-dependence of S is not an exponential but a power of t, Eq. (3.46), so that the comoving distance σ_p is an increasing function of the time of observation, not a constant. Thus points which were once at space-like distances, prohibited to exchange signals with each other, will be causally connected later, as one sees in Fig. 3.

The importance of de Sitter and anti-de Sitter spaces will become clear later when we deal with exponential expansion at very early times in inflationary scenarios.

3.4 Dynamical Parameters

In the previous section we have seen that the cosmological constant λ corresponds to a scalar, omnipresent repulsive force. If this force exactly balances the gravitational attraction on a cosmological scale, the Universe can be static, as Einstein wanted. On a small scale such as within our solar system and within our Galaxy, the attractive force in any case dominates because of the locally high density of matter. Gravitationally bound stellar systems do not expand, so we do observe that the part of the Universe comprising our Local Supercluster is static. But we know that on larger scales Hubble's law is in vigour.

Although we do not now need the cosmological constant to balance the gravitational attraction on a cosmological scale, we might still keep it in store for future uses. Perhaps a small repulsive force may be needed to match some observations. In fact, every time a new, unexplained situation arises, the cosmological constant enters the scene as a *Deus ex machina.* Let us therefore study Friedman–Lemaître–Robertson–Walker universes as given in Eqs. (3.28–29).

To characterize open, flat and closed universes we have used either the curvature parameter k with values -1, 0, +1, respectively, or the present density parameter Ω_0 with values < 1, 1, > 1, respectively. A relation between k and Ω_0 was given in Eq. (3.24) for the case when $\lambda = 0$. Let us find more general relations between the *dynamical parameters* λ, Ω_0, H_0, and the deceleration parameter q_0. Recall the definitions

$$H_0 = \frac{\dot{S}_0}{S_0} , \quad \Omega_0 = \frac{8\pi G \rho_0}{3H_0^2} , \quad q_0 = -\frac{\ddot{S}_0}{S_0 H_0^2} .$$

Substituting these into Eqs. (3.28–3.29) at present time t_0, they take the forms

$$H_0^2 + \frac{kc^2}{S_0^2} - \frac{\lambda}{3} = \Omega_0 H_0^2 \tag{3.82}$$

$$-2q_0 H_0^2 + H_0^2 + \frac{kc^2}{S_0^2} - \lambda = -3\Omega_0 H_0^2 w , \tag{3.83}$$

where w denotes the ratio $p/\rho c^2$. We can now obtain two useful relations by eliminating either k or λ. In the first case we find

$$\Omega_0(1 + 3w) = 2q_0 + \frac{2\lambda}{3H_0^2} , \tag{3.84}$$

in the second case

$$\frac{3}{2}\Omega_0(1 + w) - q_0 - 1 = \frac{kc^2}{S_0^2 H_0^2} . \tag{3.85}$$

In the present matter-dominated universe the pressure p is completely negligible, so we can set $w = 0$.

Consider a universe for which $k \geq 0$. It then follows from Eq. (3.82) that

$$\Omega_0 \geq 1 - \frac{\lambda}{3H_0^2} . \tag{3.86}$$

Thus we see that the cosmological constant changes the simple relation between k and Ω_0. For instance, a flat universe (corresponding to the equality sign) is no longer characterized by $\Omega_0 = 1$.

Note that although λ is exceedingly small, the term $\lambda/3H_0^2$ may be of the order of unity. The value $\Omega_0 = 1$ is preferred on theoretical grounds, but the observational values are mostly much smaller than 1. This is a problem we shall come back to later. Estimates of Ω_0 based on all the mass in our Local Supergalaxy yield values in the range 0.05–0.45.

When $\lambda = 0$ the solutions of the Friedman–Lemaître–Robertson–Walker equations are fully determined by the values of the three parameters t_0, H_0, Ω_0, as we have seen in the solutions (3.61) and (3.62) to the flat and open matter-dominated universes, respectively. A non-vanishing cosmological constant adds further flexibility. Whether the Universe is open or closed is then no longer determined by the value of k alone. This is explicit from Eq. (3.84) which is true independently of the value of k. In a closed Universe the expansion is always decelerated, thus $q_0 > 0$, implying

$$\Omega_0 - \frac{2\lambda}{3H_0^2} > 0 \; . \tag{3.87}$$

Thus provided Ω_0 is as small as indicated by the mass density in our Local Supergalaxy, even a moderately small positive value of λ suffices to violate this condition, making the Universe open and ever expanding.

In Fig. 17 is shown how the (Ω_0, q_0)-plane can be divided in three different ways. Universes above the line $k = 0$ have positive curvature metrics, below it they have negative curvature. Universes above the line $\lambda = 0$ develop ultimately into exponentially expanding de Sitter universes, whereas those below the line develop into contracting anti-de Sitter universes. In addition the plane can be divided by the horizontal $\Omega_0 = 1$ into overcritical universes above and undercritical below.

A recent extensive survey of all constraining astronomical evidence [4] quotes the range

$$\Omega_0 = 0.8 \pm 0.3 \; . \tag{3.88}$$

Thus it is consistent to assume that the Universe is critical, but it is not excluded that it is overcritical. This range, defined as a 68% confidence interval, is also plotted in Fig. 17. There is considerable uncertainty about the nature of matter contributing to this value

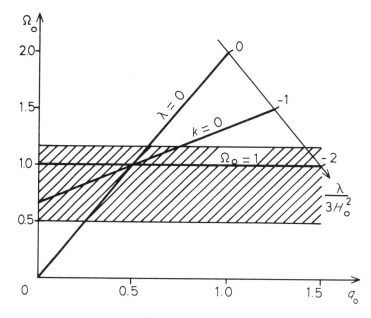

Fig. 17 Dynamical parameters in the (Ω_0, q_0) plane. Universes above (below) the line $k=0$ have positive (negative) curvature metrics. Those above (below) the line $\lambda=0$ develop into exponentially expanding de Sitter (contracting anti-de Sitter) universes. Overcritical (undercritical) universes are found above (below) the horizontal $\Omega_0 = 1$. The hatched band corresponds to the observational value of Ω_0 in Eq. (3.87)

of Ω_0. As we shall see in Section 5.3, the amount of baryonic matter falls short of this value by at least one order of magnitude. This problem may be resolved by the introduction of *dark matter* which will be discussed in Sections 7.3, 9.2 and 9.3.

To measure the deceleration parameter implies measuring a deviation from linearity in Hubble's law, as one can see from Eq. (2.16). This requires measuring the redshift and luminosity of very distant galaxies. Measurements of q_0 are hampered with difficulties which make them too imprecise at low redshifts and unreliable at high red shifts. The perhaps most reliable value comes from a survey [5] of compact radio sources which avoids the difficulties at high red shifts,

$$q_0 \simeq 0.5 , \tag{3.89}$$

which is consistent with an Einstein–de Sitter universe.

In Fig. 18 the interdependence of k, Ω_0, $\lambda/3H_0^2$ and H_0t_0 is shown. Recall that the value of H_0 is uncertain by a factor of 2 (this uncertainty was expressed by the parameter h in Table 2 on page 6), and the age of the Universe t_0 is also uncertain by a factor of about 2. However, the product H_0t_0 is better determined, so that one may take

$$H_0t_0 = 0.94 \pm 0.15 . \tag{3.90}$$

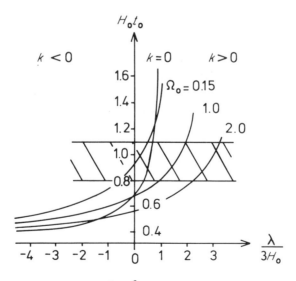

Fig. 18   Dynamical parameters in the $(\lambda/3H_0^2, H_0t_0)$ plane for a few values of Ω_0. Universes to the right (left) of the curve marked $k=0$ have positive (negative) curvature metrics. The hatched band corresponds to the observational value of H_0t_0 in Eq. (3.89). The curves are reproduced from reference [6] by permission of E. W. Kolb and the University of Chicago

From Fig. 18 one sees that a flat Einstein–de Sitter universe with $\Omega_0 = 1$ requires $\lambda = 0$ and $H_0t_0 = 2/3$. Any variation from this point can be accommodated by changing the value of at least one of the dynamical parameters. Thus we can understand the importance of obtaining precise values for the dynamical parameters: even if one limits H_0t_0 to the relatively narrow range (3.90) it is evident from Fig. 18 that one cannot at present decide whether the Universe is closed, flat or open. For a recent review of the status of the cosmological constant, see reference [8].

Problems

1. A galaxy at $z = 0.9$ contains a quasar showing redshift $z = 1.0$. Supposing that this additional redshift of the quasar is caused by its proximity to a black hole, how many Schwarzschild radii apart are they?
2. Estimate the gravitational redshift z of light escaping from a galaxy of mass $10^9 \, M_\odot$ after being emitted from the star at the radial distance 1 kpc from the centre of the galaxy[7].
3. Light is emitted horizontally *in vacuo* near the Earth's surface, and falls freely under the action of gravity. Through what vertical distances has it fallen after travelling 1 km? Calculate the radial coordinate (expressed in Schwarzschild radii) at which light travels in a circular path around a body of mass M[7].
4. On the solar surface the acceleration caused by the repulsion of a non-vanishing cosmological constant λ must be much inferior to the Newtonian attraction. Derive a value of λ from this condition.
5. Derive the expression for the age of the Universe in a closed matter-dominated Universe analogous to Eq. (3.62). This is plotted in Fig. 16 beyond the point $\Omega_0 = 1$.
6. Derive the expressions (3.72–3.73) for the time dependence of the cosmic scale factor in de Sitter models.
7. Show that the particle horizon σ_p approaches a constant also in the spatially closed de Sitter space with scale factor (3.72).
8. In Newtonian mechanics, the cosmological constant λ can be incorporated by adding to gravity an outward radial force on a body of mass m, distant r from the origin, of $F = +m\lambda r/3$. Assuming that $\lambda = -10^{-20} \, \text{yr}^{-2}$, and that F is the only force acting, estimate the maximum speed a body will attain if its orbit is comparable in size with the solar system (0.5 light-day)[7].

References

1. S. W. Hawking, *Nature*, **248** (1974) 30; *Commun. Math. Phys.*, **43** (1975) 199.
2. S. W. Hawking, *A Brief History of Time*, Bantam Books, New York, 1988.
3. E. W. Kolb and M. S. Turner, *The Early Universe*, Addison-Wesley Publ. Co., Reading, Mass., 1990.
4. N. Kaiser, G. Efstathiou, R. Ellis *et al.*, *Mon. Not. R. Astron. Soc.*, **252** (1991) 1.
5. K. I. Kellermann, *Nature*, **361** (1993) 134.
6. A. Sandage and G. A. Tammann in *Inner Space/Outer Space*, Eds. E. W. Kolb, M. S. Turner, D. Lindley, K. Olive and D. Seckel, The University of Chicago Press, Chicago, 1986, p. 41.
7. M. V. Berry, *Principles of Cosmology and Gravitation*, Adam Hilger, Bristol, 1989.
8. S. M. Carroll, W. H. Press and E. L. Turner, *Annual Review of Astronomy and Astrophysics*, **30** (1992) 499.

4 Thermodynamics

The Big Bang models describe the evolution of our Universe from a state of extreme pressure and energy density, when its size was very much smaller than now. Matter as we know it does not stand extreme temperatures. The sun is a plasma of ionized hydrogen, helium and other elements, but we know also that the stability of nuclei cannot withstand temperatures corresponding to a few MeV energy. They decompose into elementary particles which at yet higher temperatures decompose into even more elementary constituents. There is a very early time in the history of the Universe which resembles the conditions in high-energy particle colliders.

The understanding of cosmology therefore requires that we study the laws and phenomena of very high-temperature plasmas. In this chapter we shall study the thermodynamics of a plasma composed of photons, leptons, and nucleons. We now meet the electroweak force which governs the interactions of these particles during the radiation era. At energies much less than 100 GeV it is convenient to distinguish between the electromagnetic and weak interactions which are different manifestations of the electroweak force.

Motion of particles under the electromagnetic interaction is described by the Maxwell–Lorentz equations. The motion of a particle in a central field of force F, as for instance an electron of charge e moving at distance r around an almost static proton, is well approximated by the *Coulomb force*

$$F = \frac{e^2}{r^2} .$$

(4.1)

Note that this has the same form as Newton's law of gravitation, Eq. (2.3). In the electromagnetic case the strength of the interaction is e^2 whereas the strength of the gravitational interaction is GMm_G. These two *coupling constants* are expressed in completely different units because they apply to systems of completely different size. For the physics of radiation the gravitational interaction can be completely neglected, but for the dynamics of the expansion of the Universe, only the gravitational interaction is important because celestial objects are electrically neutral.

In Section 4.1 we begin with the physics of photons and Planck's radiation law which describes how the energy is distributed in an ensemble of photons in thermal equilibrium, the blackbody spectrum. We also introduce the property of spin.

In Section 4.2 we introduce the important concept of entropy and we note that a universe filled with particles and radiation in thermal equilibrium must indeed have been radiation-dominated at an early epoch. Comparing a radiation-dominated universe with one dominated by non-relativistic matter in adiabatic expansion we find that the relation between temperature and scale is different in the two cases. This leads to the conclusion that the Universe will not end in thermal death, as feared in the 19th century.

In Section 4.3 we meet new particles and antiparticles, fermions and bosons, some of their properties such as conserved quantum numbers, spin, degrees of freedom and energy spectrum, and a fair number of particle reactions describing their electroweak interactions.

In Section 4.4 we trace the thermal history of the Universe starting at a time when the temperature was 10^{13}K. The Friedmann equations offer us the means of time-keeping as a function of temperature.

In Section 4.5 we continue the thermal history from the moment of neutrino decoupling to electron decoupling to the cold microwave radiation of today.

4.1 Photons

Electromagnetic radiation in the form of radio waves, microwaves, light, X-rays or γ-rays has a dual description: either as waves characterized by the wavelength λ and frequency $\nu = c/\lambda$, or as energy quanta, *photons* γ. In the early days of quantum theory the wave–particle duality was seen as a logical paradox. It is now understood that the two descriptions are complementary, the wave picture being more useful to describe, for instance, interference phenomena, whereas the particle picture is needed to describe the kinematics of particle reactions or, for instance, the functioning of a photocell (this is what Einstein received the Nobel prize for!). Energy is not a continuous variable, but it comes in discrete packages, it is *quantized*. The quantum carried by an individual photon is

$$E = h\nu , \qquad\qquad (4.2)$$

where h is Planck's constant. The wavelength and energy ranges corresponding to the different types of radiation are given in Table 3.

Let us study the thermal history of the Universe in the Big Bang model. At the very beginning the Universe was in a state of extreme heat and pressure, occupying an exceedingly small volume. Before the onset of the present epoch, when most of the energy exists in the form of fairly cold matter, there was an era when the pressure of radiation was an important component of the energy density of the Universe, the era of *radiation domination*. As the Universe cooled, matter condensed from a hot plasma of particles and electromagnetic radiation, later to form galaxies.

During that era no atoms or atomic nuclei had yet been formed, because the temperature was too hot. Only the particles which later combined into atoms existed. These were the free electrons, protons, neutrons and various unstable particles as well as their antiparticles. Their speeds were relativistic, they were incessantly colliding and

Table 3   Electromagnetic radiation

Type	Wavelength [cm]	Energy [eV]
Radio waves	>0.1	$<10^{-3}$
Infrared	10^{-1}–7×10^{-5}	10^{-3}–1.8
Optical	$(7$–$4) \times 10^{-5}$	1.8–3.1
Ultraviolet	4×10^{-5}–10^{-6}	3.1–100
X-rays	10^{-6}–10^{-10}	100–10^{6}
γ-rays	$<10^{-10}$	$>10^{6}$

exchanging energy and momentum with each other and with the radiation photons. A few collisions were sufficient to distribute the available energy evenly among them. On average they would then have the same energy, but some particles would have less than average and some more. When the collisions resulted in a stable energy spectrum, *thermal equilibrium* was established and the photons had the *blackbody spectrum* derived in 1900 by Max Planck.

Let the number of photons of energy $h\nu$ per unit volume and frequency interval be $n_\gamma(\nu)$. Then the photon number density in the frequency interval $(\nu, \nu + d\nu)$ is

$$n_\gamma(\nu)d\nu = \frac{8\pi}{c^3} \frac{\nu^2 d\nu}{e^{h\nu/kT} - 1} . \tag{4.3}$$

At the end of the 19th century some 40 years of trial and error was spent trying to find this formula. With the hindsight of today the derivation is straightforward, based on classical thermodynamics as well as on quantum mechanics, unknown at Planck's time.

Note that Planck's formula depends on only one parameter, the temperature T. Thus the energy spectrum of photons in thermal equilibrium is completely characterized by its temperature T. The distribution (4.3) peaks at the frequency

$$\nu_{max} \simeq 6 \times 10^{10} \, T \tag{4.4}$$

in units of Hertz or cycles/s when T is given in degrees K.

The total number of photons per unit volume, or the *number density* N_γ, is found by integrating this spectrum over all frequencies,

$$N_\gamma = \int_0^\infty n_\gamma(\nu)d\nu \simeq 1.202 \frac{2}{\pi^2} \left(\frac{kT}{\hbar} \right)^3 . \tag{4.5}$$

Here $\hbar$ represents Planck's reduced constant $\hbar = h/2\pi$. The integral in this equation is not elementary, but its solution in terms of Riemann's zeta-function is well known. The occurrence of the number 1.202 is due to the properties of this function.

Since each photon of frequency ν is a quantum of energy $h\nu$ (this is the interpretation Planck was led to, much to his own dismay, because it was in obvious conflict with classical ideas of energy as a continuously distributed quantity), the total energy density of radiation is given by the *Stefan-Boltzmann law*

$$\varepsilon_\gamma = \int_0^\infty h\nu \, n_\gamma(\nu)d\nu = \frac{\pi^2}{15} \frac{k^4 T^4}{\hbar^3 c^3} \equiv aT^4, \tag{4.6}$$

where all the constants are lumped into a, called the Stefan–Boltzmann constant (see Table 2 on page 6) after Josef Stefan (1835–93) and Ludwig Boltzmann (1844–1906).

A well-known property of light is its two states of *polarization*. Unpolarized light passing a pair of polarizing sunglasses becomes vertically polarized. Unpolarized light reflected from a wet street becomes horizontally polarized. The advantage of polarizing sunglasses is that they block horizontally polarized light completely, letting all the vertically polarized light through. Their effect on a beam of unpolarized sunlight is to let on the average every second photon through vertically polarized, and to block every other photon as if it were horizontally polarized: it is absorbed in the glass. Thus the intensity of light is also reduced to one half.

In a way, it appears as if there existed two kinds of photons. Physics has taken this into account by introducing an internal property, *spin*. Thus, one can talk about the two polarization states or the two spin states of the photon.

The electromagnetic field is a vector field which at each point in space has three components, x, y and z. We can choose them as horizontal, vertical and longitudinal, the last one being along the momentum vector. However, the photon is peculiar in lacking a longitudinal polarization state. This is connected to it being massless. Recall that the theory of special relativity requires the photons to move with the speed of light in any frame. Therefore they must be massless, otherwise one would be able to accelerate them to higher speeds, or decelerate them to rest.

4.2 Adiabatic expansion

Consider a closed system of energy $E = \varepsilon V$ and pressure p contained in a comoving spherical volume $V = 4\pi S^3/3$. Let the system expand *adiabatically* under constant pressure. Recall that the *second law of thermodynamics* defines the incremental change in entropy s of particles in equilibrium at temperature T by the equation

$$ds = \frac{1}{kT}\left[d(\varepsilon V) + p dV\right], \tag{4.7}$$

where k is the Boltzmann constant. (Although it is common to denote entropy by S and entropy density by s, I reserve S for the cosmic scale factor and s for entropy.) The adiabaticity condition implies that a change in volume dV is compensated for by a change in energy dE at constant pressure and entropy, thus

$$\dot{s} = 0, \tag{4.8}$$

and in consequence Eq. (4.7) requires

$$dE = -p\, dV. \tag{4.9}$$

The second law of thermodynamics states that entropy cannot decrease in a closed system. The particles in a plasma possess maximum entropy when thermal equilibrium has been established.

Let us assume that the Universe can be treated as a non-viscous fluid, and that pressure and energy are measured by a comoving observer 'at rest' with respect to the random motion of the particles forming this fluid. The Universe then expands adiabatically and entropy is conserved. This assumption is certainly very good during the radiation-dominated era when the fluid is composed of photons and elementary particles in thermal equilibrium.

It is also true during the matter-dominated era before matter clouds start to contract into galaxies under the influence of gravity. Even on a very large scale we may consider the galaxies to form a homogeneous 'fluid', an idealization as good as the cosmological principle which forms the basis of all our discussions. In fact, we already relied on this assumption in the derivation of Einstein's equation and in the discussion of Equations of State. However, the pressure in the 'fluid' of galaxies is negligibly small, because it is caused by their random motion just as the pressure in a gas is due to the random motion of the molecules. Since the average peculiar velocities $\langle v \rangle$ of the galaxies are of the order of $10^{-3}c$, the ratio of pressure to matter density is only of the order of

$$w \approx \frac{m\langle v \rangle^2}{mc^2} = \frac{\langle v \rangle^2}{c^2} \approx 10^{-6} .$$ (4.10)

Let us compare the energy densities of radiation and matter. The energy density of electromagnetic radiation corresponding to one photon in a volume V is

$$\varepsilon_r = \frac{h\nu}{V} = \frac{hc}{V\lambda} .$$ (4.11)

In an expanding universe with cosmic scale factor S all distances scale as S, and so does the wavelength λ. The volume V then scales as S^3; thus ε_r scales as S^{-4}. Here and in the following the subscripts r and m stand for radiation and matter, respectively.

Statistical mechanics tells us that the pressure in a non-viscous fluid is related to the energy density by the equation of state

$$p = \frac{1}{3}\varepsilon$$ (4.12)

where the factor 1/3 comes from averaging over the three spatial directions. Thus pressure also scales as S^{-4}, so that it will become even more negligible in the future than it is now. The energy density of matter,

$$\varepsilon_m = \rho c^2 = \frac{Mc^2}{V}$$ (4.13)

also decreases with time, but only with the power S^{-3}. Thus the ratio of radiation energy to matter scales as S^{-1},

$$\frac{\varepsilon_r}{\varepsilon_m} \propto \frac{S^{-4}}{S^{-3}} \propto S^{-1} .$$ (4.14)

The present value of ε_r is about 10^{-12} erg/cm^3 (see Table 6 on page 112), predominantly in the form of microwaves and infrared light. The present density of matter is not known because we can only observe luminous matter (*lum*), and there may be other matter (*dark*) as well in the form of invisible particles such as neutrinos. Thus the density of luminous matter represents a lower limit,

$$\Omega_{lum} = \Omega_0 - \Omega_{dark} - \Omega_{r,0} < \Omega_0 .$$

We may then find an upper limit to the present value of the ratio of radiation to matter,

$$\frac{\varepsilon_r}{\rho_0 c^2} \simeq \frac{\Omega_{r,0}}{\Omega_0} < \frac{\Omega_{r,0}}{\Omega_{lum}} \simeq 10^{-3} . \tag{4.15}$$

Going backwards in time until the scale was smaller by this factor (light emitted then would have been redshifted by $z > 1000$) we therefore reach a time when radiation and matter contributed equally to the energy density. At that time the Universe changed from the earlier radiation-dominated era to the present matter-domination, see Fig. 19. We shall give a more precise estimate of the ratio (4.15) later.

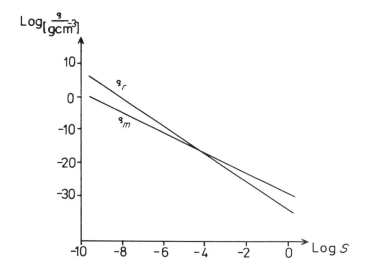

Fig. 19   Scale dependence of the energy density in radiation, ρ_r, and in matter, ρ_m

A temperature T may be converted into units of energy by the relation

$$E = kT , \tag{4.16}$$

where k is the Boltzmann constant (see Table 2 on page 6). Since E scales as S^{-1} it follows from Eq. (4.16) that also the temperature of radiation, T_r, scales as S^{-1},

$$T_r \propto S^{-1} . \tag{4.17}$$

Table 4 Particle rest masses

Particle	MeV units	K units
γ	0	0
$\nu_e, \bar{\nu}_e$	$<10^{-5}$	$<10^5$
$\nu_\mu, \bar{\nu}_\mu$	< 0.27	$<3.1\times 10^9$
$e^\pm$	0.511	5.93×10^9
$\nu_\tau, \bar{\nu}_\tau$	< 31	$<3.6\times 10^{11}$
$\mu^\pm$	105.7	1.23×10^{12}
$\pi^\pm$	139.6	1.62×10^{12}
p, n	938.3, 939.6	1.09×10^{13}
$\tau^\pm$	1784	2.07×10^{13}
$W^\pm$	80220	9.31×10^{14}
Z	91187	1.06×10^{15}

It is important to distinguish between relativistic and non-relativistic particles because their energy spectra in thermal equilibrium are different. For a particle of mass m moving with relativistic speed $\beta = v/c \lesssim 1$, the mass seen by an observer at rest is the rest mass increased by the Lorentz factor $\gamma = (1 - v^2/c^2)^{-\frac{1}{2}}$, so that it is given by

$$\frac{m}{\sqrt{1 - v^2/c^2}} \ . \tag{4.18}$$

A coarse rule is that a particle is non-relativistic when its kinetic energy is small in comparison with its rest mass, and relativistic when $E \gtrsim 10 \ mc^2$. In Table 4 the rest masses of some of the cosmologically important particles are given. For comparison, the equivalent temperatures are also given. This gives one a rough idea of the temperature of the heat bath when the respective particle is non-relativistic.

The adiabaticity condition (4.9) can be applied both to relativistic and non-relativistic particles. Let us first consider the relativistic particles dominating the radiation era. According to the theory of special relativity the energy of a particle depends on two terms, rest mass and kinetic energy. In Eq. (2.54) we already established this dependence:

$$E = \sqrt{m^2c^4 + P^2c^2} \ , \tag{4.19}$$

where P is momentum. For massless particles such as the photons, the rest mass term is absent. For any relativistic particles the mass term can also be neglected. Then Eq. (4.9) becomes

$$\mathrm{d}(S^3\varepsilon_r) = -p \ \mathrm{d}(S^3) \ . \tag{4.20}$$

Substituting ε_r for the pressure p from the equation of state (4.12) we obtain

$$S^3 \ \mathrm{d}\varepsilon_r + \varepsilon_r \ \mathrm{d}(S^3) = -\frac{1}{3}\varepsilon_r \ \mathrm{d}(S^3) \ ,$$

or

$$\frac{\mathrm{d}\varepsilon_r}{\varepsilon_r} = -\frac{4}{3}\frac{\mathrm{d}S^3}{S^3} \ . \tag{4.21}$$

The solution to this equation is

$$\varepsilon_r \propto S^{-4} \, , \tag{4.22}$$

in agreement with our previous finding. We in fact already used this result in Eq. (3.40).

For non-relativistic particles the situation is different. Their kinetic energy ε_{kin} is small, so the mass term in Eq. (4.19) can no longer be neglected. The motion of n particles per unit volume is then characterized by a temperature T_m, causing a pressure

$$p = nkT_m \, . \tag{4.23}$$

Note that T_m is not the equilibrium temperature, but rather a book-keeping device. The equation of state differs from that of radiation and relativistic matter, Eq. (4.12), by a factor 2,

$$p = \frac{2}{3} \, \varepsilon_{kin} \tag{4.24}$$

Including the rest mass term of the n particles, the energy density of nonrelativistic matter becomes

$$\varepsilon_m = nmc^2 + \frac{3}{2} nkT_m \, . \tag{4.25}$$

Substituting Eqs. (4.23) and (4.25) into the adiabaticity condition (4.20) we obtain

$$d(S^3 nmc^2) + d(S^3 \frac{3}{2} nkT_m) = -nkT_m \, dS^3 \, . \tag{4.26}$$

Let us assume that the total number of particles always remains the same: in a scattering reaction there are then always two particles coming in and two going out whatever their types. This is not strictly true because there also exist other types of reactions. However, let us assume that the total number of particles in the volume V under consideration is $N = Vn$, and that N is constant during the adiabatic expansion,

$$dN = d(Vn) = d\left(\frac{4\pi}{3} S^3 n\right) = 0 \, . \tag{4.27}$$

Then the first term in Eq. (4.26) vanishes so that we have

$$\frac{3}{2} d(S^3 T_m) = -T_m \, d(S^3) \, ,$$

or

$$\frac{3}{2} \frac{dT_m}{T_m} = -\frac{d(S^3)}{S^3} \, .$$

The solution to this differential equation is of the form

$$T_m \propto S^{-2} \ . \tag{4.28}$$

Thus we see that the temperature of non-relativistic matter has a different dependence on the scale of expansion than does the temperature of radiation. This has profound implications for one of the most serious problems in thermodynamics in the 19th century.

Suppose that the Universe starts out at some time with γ rays at high energy and electrons at rest. This would be a highly ordered non-equilibrium system. The photons would obviously quickly distribute some of their energy to the electrons via various scattering interactions. Thus the original order would decrease, and the randomness or disorder would increase. The *second law of thermodynamics* states that any isolated system left by itself can only change towards greater disorder. The measure of disorder is *entropy*; thus the law says that entropy cannot decrease.

The counter-example which living organisms seem to furnish since they build up ordered systems is not a valid counter-example. This is so because no living organism exists in isolation; it consumes nutrients and produces waste. Thus to establish that a living organism indeed increases entropy would require measuring a much larger system, certainly not smaller than the solar system.

It now seems to follow from the second law of thermodynamics that all energy would ultimately distribute itself evenly throughout the Universe, so that no temperature differences would exist anymore. The discoverer of the law of conservation of energy, Hermann von Helmholtz (1821–1894), came to the distressing conclusion in 1854 that 'from this point on, the Universe will be falling into a state of eternal rest'. This state was named the *thermal death*, and it preoccupied greatly both philosophers and scientists during the 19th century.

Now we see that this pessimistic conclusion was premature. Because, if there was a time when the temperatures of matter and radiation were the same,

$$T_m = T_r \ ,$$

we see from Eqs. (4.17) and (4.28) that the expansion of the Universe will cause matter to cool faster than radiation. Thus cold matter and hot radiation in an expanding Universe are not and will never be in thermal equilibrium on a cosmic time scale. This result permits us to solve the adiabatic equations of cold matter and hot radiation separately, as we in fact did.

4.3 Electroweak Interactions

In *quantum electro-dynamics* (QED) the electromagnetic field is mediated by photons which are emitted by one charged particle and absorbed very shortly afterwards by another. Such photons with a brief existence during an interaction are called *virtual*, in contrast to real photons.

Virtual particles do not travel freely to or from the interaction region. In the production of virtual particles energy is not conserved. This is possible because the energy imbalance arising at the creation of the virtual particle is compensated for when it is annihilated, so that the real particles emerging from the interaction region possess the same amount of energy as those entering the region. We have already met this argument in the discussion of Hawking radiation from black holes.

However, Nature impedes the creation of very huge energy imbalances. For example, the masses of the *vector bosons* $W^{\pm}$ and Z^0 mediating the electroweak interactions are almost 100 GeV. Reactions at much lower energies involving virtual vector bosons are therefore severely impeded, and much less frequent than electromagnetic interactions. For this reason such interactions are called *weak interactions*.

Real photons interact only with other charged particles such as protons p, electrons e^- and their oppositely charged *antiparticles*, the *antiproton* $\bar{p}$ and the *positron* e^+. An example is the elastic *Compton scattering* of electrons by photons,

$$\gamma + e^{\pm} \rightarrow \gamma + e^{\pm} \ . \tag{4.29}$$

As an effect of virtual intermediate states neutral particles may exhibit electromagnetic properties such as magnetic moment.

Antimatter does not exist on Earth, and there is very little evidence for its presence elsewhere in the Galaxy. That does not mean that antiparticles are pure fiction: they are readily produced in particle accelerators and in violent astrophysical events. However, in an environment of matter antiparticles rapidly meet their corresponding particles and annihilate each other.

Charged particles interact via the electromagnetic field. Examples are the elastic scattering of electrons and positrons,

$$e^{\pm} + e^{\pm} \rightarrow e^{\pm} + e^{\pm} \ , \tag{4.30}$$

and the Coulomb interaction between an electron and a proton, depicted in Fig. 20. The free e^- and p enter the interaction region from the left, time running from left to right. They then exchange a virtual photon, and finally they leave the interaction region as free particles. This *Feynman diagram* does not show that the energies and momenta of the e^- and p change in the interaction. If one particle is fast and the other slow, the result of the interaction is that the slow particle picks up energy from the fast one, just as in the case of classical billiard balls. The Coulomb interaction between particles of like charges is repulsive, between unlike charges attractive. In both cases the energy and momentum get redistributed in the same way.

When an electron is captured by a free proton, they form a bound state, a hydrogen atom which is a very stable system. An electron and a positron may also form a bound atom-like state called *positronium*. This is a very unstable system: the electron and positron are antiparticles so they rapidly end up annihilating each other according to the reaction

$$e^- + e^+ \rightarrow \gamma + \gamma \ . \tag{4.31}$$

Since the total energy is conserved, the annihilation results in two (or three) photons possessing all the energy and flying away with it at the speed of light.

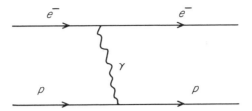

Fig. 20 Feynman diagram for elastic scattering of an electron e^- on a proton p. This is an electromagnetic interaction mediated by a virtual photon γ. The direction of time is from left to right

The reverse reaction is also possible. A photon may convert briefly into a virtual $e^- e^+$ pair, and another photon may collide with either one of these charged particles, knocking them out of the virtual state, thus creating a free electron–positron pair

$$\gamma + \gamma \rightarrow e^- + e^+ \,. \tag{4.32}$$

This requires the energy of each photon to equal at least the electron and positron rest mass, 0.51 MeV. If the photon energy is in excess of 0.51 MeV the $e^- e^+$ pair will not be created at rest, but both particles will acquire kinetic energy.

Protons and antiprotons have electromagnetic interactions similar to positrons and electrons. They can also annihilate into photons, or for instance into an electron–positron pair via the mediation of a virtual photon

$$p + \bar{p} \longrightarrow \gamma_{virtual} \rightarrow e^- + e^+ \,, \tag{4.33}$$

as depicted in Fig. 21. The reverse reaction

$$e^- + e^+ \longrightarrow \gamma_{virtual} \rightarrow p + \bar{p} \tag{4.34}$$

is also possible provided the electron and positron possess enough kinetic energy to create a proton, or 938.3 MeV.

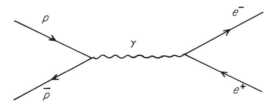

Fig. 21 Feynman diagram for $p\bar{p}$ annihilation into $e^+ e^-$ via an intermediate virtual photon γ. The direction of time is from left to right

Note that the total electric charge is conserved throughout the reactions (4.29–4.34) and in Figs. 20–21. Its value after the interaction (to the right of the arrow) is the same as it was before (to the left of the arrow). This is an important conservation law: electric charge can never disappear nor arise out of neutral vacuum. In the annihilation of an $e^- e^+$ pair into photons, all charges do indeed vanish, but only because the sum of the charges was zero to start with.

All the charged particles mentioned have neutral partners as well. The partners of the p, $\bar{p}$, e^-, e^+ are the *neutron n*, the *antineutron* $\bar{n}$, the *electron neutrino* ν_e and the *electron antineutrino* $\bar{\nu}_e$. The p and n are called *nucleons*, they belong together with a host of excited nucleon-like states to the more general family of *baryons*. The $\bar{p}$ and $\bar{n}$ are correspondingly *antinucleons* or *antibaryons*. The e^-, e^+, ν_e and $\bar{\nu}_e$ are called *leptons* and *antileptons* of the *electron family (e)*.

We also have to introduce two more families of leptons: the μ family comprising the charged *muons* $\mu^{\pm}$ and their associated neutrinos ν_{μ}, $\bar{\nu}_{\mu}$, and the τ family comprising $\tau^{\pm}$ and ν_{τ}, $\bar{\nu}_{\tau}$. The $\mu^{\pm}$ and $\tau^{\pm}$ are much more massive than the electrons, but otherwise their physics is very similar. They participate in reactions such as Eqs. (4.29–4.32) with e replaced by μ or τ, respectively.

The charge can easily move from a charged particle to a neutral one as long as that does not violate the conservation of total charge in the reaction. We need to know two further conservation laws governing the behaviour of baryons and leptons:

(i) *B* or *baryon number* is conserved. This forbids the total number of baryons minus antibaryons to change in particle reactions. To help the book-keeping in particle reactions we assign the value $B = 1$ to baryons and $B = -1$ to antibaryons in a way analogous to the assignment of electric charges. Photons and leptons have $B = 0$.

(ii) L_l or *l-lepton number* is conserved for each of the families $l = e, \mu, \tau$. This forbids the total number of *l*-leptons minus $\bar{l}$-antileptons to change in particle reactions. We assign $L_e = 1$ to e^- and ν_e, $L_e = -1$ to e^+ and $\bar{\nu}_e$, and correspondingly to the members of the μ and τ families. Photons and baryons have no lepton numbers.

All leptons participate in the weak interactions mediated by the heavy virtual vector bosons $W^{\pm}$ and Z^0. The Z^0 is just like a photon except that it is very massive, about 91 GeV, and the $W^{\pm}$ are its 10 GeV lighter charged partners. Weak leptonic reactions are

$$e^{\pm} + \overset{(-)}{\nu}_e \rightarrow e^{\pm} + \overset{(-)}{\nu}_e \, , \tag{4.35}$$

$$\overset{(-)}{\nu}_e + \overset{(-)}{\nu}_e \rightarrow \overset{(-)}{\nu}_e + \overset{(-)}{\nu}_e \, , \tag{4.36}$$

where $\overset{(-)}{\nu}_e$ stands for ν_e or $\bar{\nu}_e$. In Fig. 22 the Feynman diagrams of some of these reactions are shown. There is also the annihilation reaction

$$e^- + e^+ \rightarrow \nu_e + \bar{\nu}_e \, , \tag{4.37}$$

and the pair production reaction

$$\nu_e + \bar{\nu}_e \rightarrow e^- + e^+ \, . \tag{4.38}$$

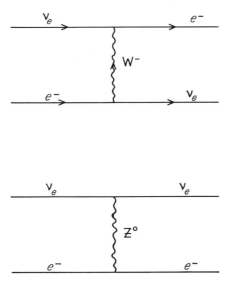

Fig. 22 Feynman diagram for elastic scattering of an electron neutrino ν_e against an electron e^-. This weak interaction is mediated by a virtual W^- vector boson in the charged current reaction (upper figure), and by a virtual Z^0 vector boson in the neutral current reaction (lower figure). The direction of time is from left to right

Similar reactions apply to the two other lepton families, replacing e above by μ or τ, respectively. Note that the total baryon number B and the total lepton number L_e are both conserved throughout the above reactions. Figure 22 illustrates that the ν_e can scatter against electrons by the two Feynman diagrams corresponding to $W^\pm$ exchange and Z^0 exchange, respectively. In contrast, ν_μ and ν_τ can only scatter by the Z^0 exchange diagram, because of the separate conservation of lepton-family numbers.

The leptons and nucleons all have two spin states each. In the following we shall refer to them as *fermions*, whereas the photon and the W and Z are *bosons*. In Table 4 (page 75) we have already seen one more boson, the π meson, or *pion*. The difference between bosons and fermions is deep and fundamental. The number of spin states is even for fermions, odd for bosons (except the photon). They behave differently in a statistical ensemble. Fermions have antiparticles which most bosons do not. The *fermion number* is conserved, indeed separately for leptons and baryons, as we have seen. The number of bosons is not conserved, for instance in pp collisions one can produce any number of pions and photons.

Two identical fermions refuse to get too close to one another. This is the *Pauli exclusion force* responsible for the electron *degeneracy pressure* in white dwarfs and the neutron degeneracy pressure in neutron stars. A gas of free electrons will exhibit pressure even at temperature of absolute zero. According to quantum mechanics, particles never have exactly zero velocity, they always carry out random motions causing pressure. For electrons in a high-density medium such as a white dwarf with density $10^6 \rho_\odot$, the degeneracy pressure is much larger than the thermal pressure, and it is enough to balance the pressure of gravity.

Bosons do not feel such a force, nothing inhibiting them to get close to each other. However, it is beyond the scope and needs of this book to explain these properties further. They belong to the domains of quantum mechanics and quantum statistics.

The massive vector bosons $W^\pm$ and Z^0 have three spin or polarization states: the *transversal* (vertical and horizontal) states which the photons also have, and the *longitudinal state* along the direction of motion which the photon is lacking.

Table 5   Particle degrees of freedom

Particle	Particle type	n_{spin}	n_{anti}	g
γ	vector boson	2	1	2
ν_e, ν_μ, ν_τ	fermion (lepton)	1	2(?)	$\frac{7}{4}$(?)
e^-, μ^-, τ^-	fermion (lepton)	2	2	$\frac{7}{2}$
$\pi^\pm, \pi^0$	boson (meson)	1	1	1
p, n	fermion (baryon)	2	2	$\frac{7}{2}$
$W^\pm, Z$	vector boson	3	1	3

In Table 5 the number of spin states, n_{spin}, of some of the cosmologically important particles are given (note that neutrinos have only half the number of spin states than other fermions). The fourth column tabulates n_{anti}, which equals 2 for particles which possess a distinct antiparticle, otherwise 1. For the neutrinos this information is still uncertain.

As already explained, the number of distinct states or *degrees of freedom*, g, of photons in a statistical ensemble (in a plasma, say) is 2. In general, due to the intricacies of quantum statistics, the degrees of freedom are the product

$$g = n_{spin}\, n_{anti}\, n_{Pauli}, \tag{4.39}$$

where $n_{Pauli} = \frac{7}{8}$ for fermions and 1 for bosons. This product is tabulated in the fifth column of Table 5.

The *rate* at which a reaction occurs, or the number of events per unit time, depends on the strength of the interaction as expressed by the coupling constant. It may also depend on many other details of the reaction such as the spins and masses of the participating particles, and the energy E. All this information is contained in the reaction *cross section*, σ. Let us follow an elementary argument to derive this quantity.

Suppose a beam contains k monoenergetic particles per cm³, all flying with velocity v cm/s in the same direction, see Fig. 23. This defines the *flux* F of particles per cm²s in the beam. Let the beam hit a surface containing N target particles on which each beam particle may scatter. The number of particle reactions (actual scatterings) per second is then proportional to F and N. Consider the number of particles dN_{scat} scattered into a detector of size $d\Omega$ in a direction θ from the beam direction (we assume azimuthal symmetry around the beam direction). Obviously dN_{scat} is proportional to the number of particle reactions and to the detector opening,

$$dN_{scat} = FN\sigma(\theta)d\Omega \ .$$

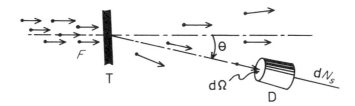

Fig. 23 A beam of particles of flux F hitting a target T which scatters some of them into a detector D in the direction θ. The detector, which has a sensitive surface $d\Omega$, then records dN_s scattered particles

The proportionality factor $\sigma(\theta)$ contains all the detailed information about the interaction. Integrating over all directions we can write this

$$N_{scat} = FN\sigma \ , \tag{4.40}$$

where the proportionality constant

$$\sigma \equiv \int \sigma(\theta) d\Omega$$

has the dimension of a surface, here cm^2. For this reason it has been named *cross section*.

One can also understand the reason for the surface units from a classical argument. Suppose a particle reaction can be treated as a game of billiard balls. Then the probability of hit is clearly proportional to the size of the target ball (of radius R), as seen by the hitting ball, or πR^2. The difference between billiard balls and particles is that σ should not be understood as the actual size, because that is not a useful quantity in quantum mechanics. Rather it depends on the interaction in a complicated manner.

The number density of other relativistic particles than photons is given by distributions very similar to the Planck distribution. Let us replace the photon energy $h\nu$ in Eq. (4.3) by E which is given by the relativistic expression (4.19). Noting that the kinematic variable is now the three-momentum $p = |\mathbf{p}|$, we can replace Planck's distribution by the number density of particle species i with momentum between p and $p+dp$,

$$n_i(p)dp = \frac{8\pi}{h^3} \frac{n_{spin,i}}{2} \frac{p^2 dp}{e^{E_i(p)/kT_i} \pm 1} \ . \tag{4.41}$$

The $\pm$ sign is – for bosons and + for fermions, and the name for these distributions are the *Bose distribution* and the *Fermi distribution*, respectively. The Fermi distribution in the above form is actually a special case: it holds when the number of charged fermions equals the number of corresponding neutral fermions (the 'chemical potentials' vanish). In the following we shall need only that case.

The number density N of non-relativistic particles of mass m is given by the *Maxwell–Boltzmann* distribution for an ideal, non-degenerate gas. Starting from Eq. (4.41) we note that for nonrelativistic particles the energy kT is smaller than the rest mass, so that the term ± 1 in can be neglected in comparison with the exponential. Rewriting the

Fermi distribution as a function of temperature rather than of momentum we obtain the Maxwell–Boltzmann distribution

$$N = n_{spin} \frac{(2\pi mkT)^{3/2}}{(ch)^3} \; \mathrm{e}^{-E_i/kT_i} \;.$$ (4.42)

James Clerk-Maxwell (1831–79) was a contemporary of Stefan and Boltzmann.

4.4 The Early Radiation Era

In Section 4.1 we established the dependence of the number density of photons on temperature, N_γ in Eq. (4.5), and the corresponding energy density, ρ_γ in Eq. (4.6). For each species of relativistic fermions participating in the thermal equilibrium there is a specific number density.

To find the total number density of particles sharing the available energy we have to count each particle species i weighted by the corresponding degrees of freedom g_i. Remembering that $g_\gamma = 2$ for photons, we rewrite Eq. (4.6) with a factor g_i explicitly visible, thus

$$\varepsilon_i = \frac{g_i}{2} aT^4 \;.$$ (4.43)

It turns out that this expression gives the correct energy density for every particle species if we insert its respective value of g_i from Table 5 (page 82).

The equation (4.5) can be correspondingly generalized to relativistic fermions. Their number density is

$$N_f = \frac{3}{4} N_\gamma \;.$$ (4.44)

In general, the primordial plasma is a mixture of particles of which some are relativistic and some non-relativistic at a given temperature. Different species may also have a thermal distribution with a different temperature than the photons. Let us for the moment ignore such differences, and define the *effective degrees of freedom* of the mixture by

$$g_{eff} = \sum_i g_i \;.$$ (4.45)

Thus the energy density of the mixture is

$$\varepsilon_r = \frac{g_{eff}}{2} aT^4 \;.$$ (4.46)

Let us now derive a relation between the temperature scale and the time scale. We have already found the relation (3.48) between the size scale S and the time scale t during the radiation era,

$$S(t) \propto \sqrt{t} \;,$$ (4.47)

where we choose to omit the proportionality factor. Note that this dependence was derived assuming that time started at $t_{start} = 0$ and at scale $S = 0$. We shall now be interested in the time elapsed between different epochs, so we must reintroduce the integration constant neglected in Eq. (3.48). Then t is replaced by $t - t_{start}$. The Hubble parameter can then be written

$$H = \frac{\dot{S}}{S} = \frac{1}{2(t - t_{start})} .$$
(4.48)

Note that the proportionality factor omitted in Eq. (4.47) has dropped out.

In Eq. (3.43) we noted that the curvature term kc^2/S^2 in Friedman's equations is negligibly small at early times during the radiation era. We then obtained the dynamical relation

$$\frac{\dot{S}}{S} = \left(\frac{8\pi}{3}G\rho\right)^{\frac{1}{2}} .$$
(4.49)

Inserting Eq. (4.48) on the left and replacing the energy density ρ on the right by ε_r/c^2, we find the sought relation between photon temperature and time,

$$\frac{1}{t - t_{start}} = \sqrt{\frac{16\pi Ga}{3c^2}g_{eff}}\ T^2 = 3.07 \times 10^{-21} \sqrt{g_{eff}} \frac{T^2}{[K^2]}[s^{-1}] .$$
(4.50)

Let us now study the thermal history of the Universe during the radiation era. We may start at a time when the temperature was 10^{13} K which corresponds to about 860 MeV. This is just below the threshold for proton–antiproton production, Eq. (4.46). Thus the number of protons (and neutrons and antinucleons as well) will no longer increase as a result of thermal collisions. They can only decrease for a number of reasons which I shall explain. Actually, the number of nucleons must then have been quite small already, because today there exists only one nucleon for each $3 - 4 \times 10^{10}$ photons.

Most of the other particles introduced in Section 4.3, the $\gamma, e^-, \mu^-, \pi, \nu_e, \nu_\mu, \nu_\tau$ as well as their antiparticles are then present. The sum in Eq. (4.45) is then, using the degrees of freedom in Table 5 (page 83),

$$g_{eff} = 2 + 3 + 2 \times \frac{7}{2} + 3 \times \frac{7}{4} = \frac{69}{4} ,$$
(4.51)

where the first term corresponds to the photon, the second to the three pions, the third to the two charged leptons and the fourth to the three kinds of neutrinos. The unstable τ leptons have already disappeared shortly after 1.78 GeV because their mean life is of the order of picoseconds. Inserting the value of g_{eff} into Eq. (4.50), and choosing time $t = 0$ at 10^{13} K, we find that $t_{start} = -0.66$ μs.

At this time the number density of nucleons is decreasing fast because they have become non-relativistic. In consequence, they have a larger probability to annihilate into lepton pairs, pion pairs or photons. Their number density is then no longer given by the Fermi distribution (4.41), but by the Maxwell–Boltzmann distribution, Eq. (4.42). As can be seen from the latter, when T drops below the rest mass the number density decreases rapidly because of the exponential factor. If there had been exactly the same number of nucleons and antinucleons, we would not expect many nucleons to remain to form matter. But, since we live in a matter universe, there must have been some excess of nucleons early on. Note that neutrons and protons exist in equal numbers at the time under consideration.

Although the nucleons are very few, they still participate in electromagnetic reactions such as the elastic scattering of electrons,

$$e^{\pm} + p \rightarrow e^{\pm} + p \,, \tag{4.52}$$

and in weak *charged current* reactions in which charged leptons and nucleons change into their neutral partners, and vice versa, as in

$$e^{-} + p \rightarrow \nu_e + n \,, \tag{4.53}$$

$$\bar{\nu}_e + p \rightarrow e^{+} + n \,. \tag{4.54}$$

Other such reactions are obtained by reversing the arrows, and by replacing $e^{\pm}$ by $\mu^{\pm}$ or ν_e by ν_μ or ν_τ. The nucleons still participate in thermal equilibrium, but they are too few to play any rôle in the thermal history anymore. This is why we could neglect them in Eq. (4.51).

Below the pion rest mass (actually at about 70 MeV) the temperature in the Universe cools below the threshold for pion production,

$$(e^{-} + e^{+}) \text{ or } (\mu^{-} + \mu^{+}) \rightarrow \gamma_{virtual} \rightarrow \pi^{+} + \pi^{-} \,. \tag{4.55}$$

The reversed reactions, pion annihilation, still operate, reducing the number of pions. However, they disappear even faster by decay. This is always the fate when such lighter states are available, that energy and momentum are conserved, as well as quantum numbers such as electric charge, baryon number and lepton numbers. The pion, the muon and the tau lepton are examples of this. The pion decays mainly by the reactions

$$\pi^{-} \rightarrow \mu^{-} + \bar{\nu}_\mu \,, \quad \pi^{+} \rightarrow \mu^{+} + \nu_\mu \,. \tag{4.56}$$

Thus g_{eff} decreases by 3 to 57/4. The difference in rest mass between the initial pion and the final state particles is

$$m_\pi - m_\mu - m_\nu = (139.6 - 105.7 - 0.0) \text{ MeV} = 33.9 \text{MeV} \,, \tag{4.57}$$

so 33.9 MeV is available as kinetic energy to the muon and the neutrino. This makes it very easy for the $\pi^{\pm}$ to decay, and in consequence its mean life is short, only 0.026 μs (the π^0 decays even much faster). This is much less than the age of the Universe at 140 MeV which is 27 μs from Eq. (4.50). (This value is a bit too small because a detailed calculation requires integrating over the smooth function $g_{eff}(T)$ which is not quite a stepfunction, see Fig. 24 [1]). Note that the appearance of a charged lepton in the final state forces the simultaneous appearance of its antineutrino in order to conserve lepton number.

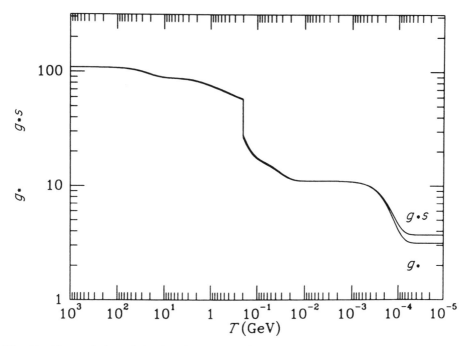

Fig. 24 The evolution of the function $g_* = 4g_{eff}(T)$ as a function of temperature T, reproduced from reference [1] by permission of E.W. Kolb and M. Turner. When the effective degrees of freedom are needed to evaluate the entropy in the matter-dominated epoch, g_* has to be replaced by g_{*S}

Also the muons decay fast compared to the age of the Universe, with a mean life of 2.2 μs, by the processes

$$\mu^- \rightarrow e^- + \bar{\nu}_e + \nu_\mu , \quad \mu^+ \rightarrow e^+ + \nu_e + \bar{\nu}_\mu . \quad (4.58)$$

Almost the entire rest mass of the muon, or 105.7 MeV is available as kinetic energy to the final state particles. This is the reason for its short mean life. Here again the conservation of lepton numbers, separately for the e-family and the μ-family, is observed.

Below the muon rest mass (actually at about 50 MeV) the temperature in the Universe cools below the threshold for muon pair production,

$$e^- + e^+ \rightarrow \gamma_{virtual} \rightarrow \mu^+ + \mu^- . \quad (4.59)$$

The time elapsed is less than a millisecond. When the muons have disappeared, we can reduce g_{eff} by 7/2 to 43/4.

From the reactions (4.56) and (4.58) we see that the end products of pion and muon decay are stable electrons and neutrinos. The electron neutrino is certainly stable; whether also the muon and tau neutrinos are stable we do not know at this time. We are now left with neutrinos which only participate in weak reactions, and with electrons, photons and a very small number of nucleons. The number density of each lepton species is about the same as that of photons.

4.5 The Time of Decoupling

The considerations about which particles participate in thermal equilibrium at a given time depend on two time scales: the *reaction rate* of the particle, taking into account what reactions are possible at that energy, and the *expansion rate* of the Universe. If the reaction rate is slow compared to the expansion rate, the distance between particles grows so fast that they cannot find each other.

The expansion rate is given by $\dot{S}/S$ which is the same as the Hubble parameter H. From Eqs. (4.48) and (4.50) its temperature dependence is

$$ H = \frac{\dot{S}}{S} = \sqrt{\frac{16\pi Ga}{3c^2} g_{eff}} \; T^2 \; . \tag{4.60} $$

The average reaction rate can be written

$$ \Gamma = \langle N v \sigma(E) \rangle \; , \tag{4.61} $$

where $\sigma(E)$ is the reaction cross section (in units of cm^2) as defined in Eq. (4.40). The product of $\sigma(E)$ and the velocity v of the particle varies over the thermal distribution, so one has to average over it, as is indicated by the angular brackets. Multiplying this product by the number density N of particles per cm^3 one obtains the mean rate Γ of reacting particles per second, or the mean free path between collisions, Γ^{-1}.

The weak interaction cross section turns out to be proportional to T^2,

$$ \sigma \simeq \frac{G_F^2 (kT)^2}{\pi (\hbar c)^4} , \tag{4.62} $$

where G_F is the *Fermi coupling* measuring the strength of the weak interaction. The number density of the neutrinos is proportional to T^3 according to Eq. (4.5). The reaction rate of neutrinos then falls with decreasing temperature as T^5.

The condition for a given species of particle to remain in thermal equilibrium is then that the reaction rate Γ is larger than the expansion rate H, or equivalently that Γ^{-1} does not exceed the Hubble distance H^{-1},

$$ \frac{\Gamma}{H} \gtrsim 1 \; . \tag{4.63} $$

Inserting the T^5-dependence of the weak interaction interaction rate Γ_{wi} and the T^2-dependence of the expansion rate H from Eq. (4.60) we obtain

$$ \frac{\Gamma_{wi}}{H} \propto T^3 \; . \tag{4.64} $$

Thus there may be a temperature small enough so that the condition (4.63) is no longer fulfilled. The neutrinos then *decouple* or *freeze out* from all interactions, and begin a free expansion. The decoupling of ν_μ and ν_τ occurs at 3.5 MeV, whereas the ν_e decouple at 2.3 MeV. This can be depicted as a set of connecting baths containing different particles, and having valves which close at given temperatures, see Fig. 25.

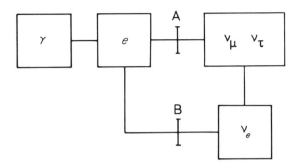

Fig. 25 A system of communicating vessels illustrating particles in thermal equilibrium. At 3.7 MeV the valve A closes so that ν_μ and ν_τ decouple. At 2.3 MeV the valve B closes so that also ν_e decouples, leaving only e^- and γ in thermal contact

At decoupling, the neutrinos are still relativistic since they are either massless or very light (see Table 3, p. 71). Thus their energy distribution is given by the Fermi distribution, Eq. (4.41), and their average temperature T_ν equals that of the photons. The neutrino degrees of freedom continue to contribute 21/4 to g_{eff} even when they have lost thermal contact with the photon plasma. After decoupling T_ν continues to decrease with the increasing size of the Universe as S^{-4}.

Let us now turn to the fate of the remaining nucleons. Note that the charged current reactions (4.53–4.54) changing a proton to a neutron are *endothermic*. They require some input energy to provide for the mass difference. In reaction (4.53) this difference is 0.8 MeV, in reaction (4.54) it is 1.8 MeV (use the rest masses in Table 4, p. 75!). The reversed reactions are *exothermic*. They liberate energy so that they can always proceed without any threshold limitation.

As the Universe cools and the energy approaches 0.8 MeV, the endothermic neutron-producing reactions stop, one by one. Thus, no more neutrons are produced but some of them get converted into protons in the exothermic reactions. Thus the ratio of protons to neutrons increases, ultimately reaching the value 5.

The neutrons also decay into protons by *beta decay*,

$$n \rightarrow e^- + \bar{\nu}_e + p \, , \tag{4.65}$$

liberating $m_n - m_p - m_e - m_\nu = 0.8$ MeV kinetic energy in the process. This amount is very small compared to the neutron rest mass of 939.6 MeV. In consequence the decay is inhibited and very slow: the neutron mean life is 889 s. In comparison with the age of the Universe which at this time is a few tens of seconds the neutrons are essentially stable. The protons are stable even on scales of billions of years, so their number is not going to decrease by decay.

Photons with energies below the electron rest mass can no longer produce e^+–e^- pairs, but the energy exchange between photons and electrons still continues by Compton scattering, reaction (4.29), or *Thomson scattering* as it is called at very low energies. Electromagnetic cross sections (subscript *em*) are proportional to T^{-2}, and the reaction rate is then proportional to T so that

$$\frac{\Gamma_{em}}{H} \propto \frac{1}{T} .$$

Contrary to the weak interaction case, Eq. (4.64), the condition (4.63) is then satisfied for all temperatures, so electromagnetic interactions never freeze out.

The exothermic electron-positron annihilation, reaction (4.31), is now of mounting importance, creating new photons with energy 0.51 MeV. This is higher than the ambient photon temperature at that time, so the photon population gets reheated. To see just how important this reheating is, let us turn to the law of conservation of entropy.

By integrating the right-hand side of Eq. (4.7) over V one finds that entropy can be defined by

$$s = \frac{V}{kT} \ (\rho c^2 + p) . \tag{4.66}$$

Making use of the Equation of State for relativistic particles (4.12), this can be written

$$s = \frac{4V}{3kT} \ \varepsilon_r . \tag{4.67}$$

Substituting the expression for ε_r from Eq. (4.46) we have

$$s = \frac{2}{3} \ g_{eff} \frac{VaT^4}{kT} . \tag{4.68}$$

Now aT^4 is energy density so VaT^4 is energy, and kT is also energy, so VaT^4/kT is just a constant. Also g_{eff} is a constant, except at the thresholds where particle species decouple. Hence between two decoupling thresholds the entropy in a comoving volume V is a constant, and conserved as assumed,

$$\frac{ds}{dt} = \frac{d}{dt} \left(\frac{2}{3} \ g_{eff} \frac{VaT^3}{k} \right) = 0 . \tag{4.69}$$

The second law of thermodynamics requires that entropy should be conserved in reversible processes, also at thresholds where g_{eff} changes. This is only possible if T also changes in such a way that $g_{eff}T^3$ remains a constant. When a relativistic particle becomes non-relativistic and disappears, its entropy is shared between the particles remaining in thermal contact, thus causing some slight slowdown in the cooling rate. Photons and massless neutrinos never become non-relativistic, therefore they continue to share the entropy of the Universe, each species conserving its entropy separately.

Let us now apply this argument to the situation when the positrons and most of the electrons disappear by annihilation below 0.2 MeV. We denote quantities just above this energy by a subscript +, and below it by –. Above this energy the particles in thermal equilibrium are the γ, e^-, e^+. Then the entropy is

$$s = \frac{2}{3}(2 + \frac{7}{2})\frac{VaT_+^3}{k} . \tag{4.70}$$

Below that energy only photons contribute the factor $g_{eff} = 2$ anymore. In consequence, the ratio of entropies s_+ and s_- is

$$\frac{s_+}{s_-} = \frac{11}{4}\left(\frac{T_+}{T_-}\right)^3 .$$ (4.71)

But entropy must be conserved so this ratio must be 1. It then follows that

$$T_- = \left(\frac{11}{4}\right)^{\frac{1}{3}} T_+ = 1.40\, T_+ .$$ (4.72)

Thus the temperature T_γ of the photons increases by a factor 1.40 as the Universe cools below the threshold for electron–positron pair production. Actually the temperature increase is so small and so gradual that it only slows down the cooling rate temporarily.

Up to this moment the decoupled neutrinos have had the same temperature as the photons, $T_\nu = T_\gamma$. But the neutrinos do not participate in the reheating process, and they do not share the entropy of the photons, so that from now on they are colder than the photons,

$$T_\nu = T_\gamma/1.40 .$$ (4.73)

The number density N_ν of neutrinos can be calculated as in Eq. (4.5) using Eq. (4.3), except that the -1 term in Eq. (4.3) has to be replaced by $+1$ as is required for fermions, see Eq. (4.41). The result is that N_ν is a factor 3/4 times N_γ at the same temperature. Taking the different of temperatures T_ν and T_γ into account and noting from Eq. (4.5) that N_γ is proportional to T^3, one finds

$$N_\nu = \frac{3}{4}\cdot\frac{4}{11}N_\gamma.$$ (4.74)

The physical meaning of entropy of a system is really its degrees of freedom times some constant, as one sees from Eq. (4.69). For instance, the entropy density of photons can be written

$$\frac{s}{V} = 1.80\, g_{eff}\, N_\gamma ,$$ (4.75)

where N_γ is the number density of photons. The sum of degrees of freedom of a system of particles is of course the number of particles multiplied by the degrees of freedom per particle. Independently of the law of conservation of energy, the conservation of entropy implies that the energy is distributed equally between all degrees of freedom present in such a way that a change in degrees of freedom is accompanied by a change in random motion, or equivalently in temperature.

Thus entropy is related to order: the more degrees of freedom there are present, the more randomness or disorder does the system possess. When an assembly of particles (such as the molecules in a gas) does not possess other energy than kinetic energy (heat), its entropy is maximal when thermal equilibrium is reached. For a system of gravitating bodies entropy increases by clumping, maximal entropy corresponding to a black hole.

To end this chapter on photons and leptons, let us still follow their fate, leaving for later the important events in which the nucleons participate. As long as there are free electrons, the photons will scatter against them (Thomson scattering). The reason why the slow, non-relativistic electrons do not decouple as did the neutrinos, is that they interact with photons. Thus the interaction is electromagnetic, so the reaction rate is much higher than the weak reaction rate of the neutrinos. In fact, the reaction rate is higher than the expansion rate of the Universe, so the condition (4.63) is fulfilled.

Since the photons continue to meet electrons, they do not propagate in straight paths for very long distances. In other words, the Universe is opaque to electromagnetic radiation including light. It would have been impossible to do astronomy if this situation would have persisted until our days.

When the electrons become slow enough, they are captured into atomic orbits by the protons, forming stable hydrogen atoms and other light atoms. Unlike the unstable particles n, π, μ which decay spontaneously liberating kinetic energy in exothermic reactions, the hydrogen atom H is a *bound state* of a proton and an electron. Its mass is less than the p and e^- rest masses together,

$$m_{\mathrm{H}} - m_p - m_e = -13.59 \text{ eV} \qquad (4.76)$$

so it cannot disintegrate spontaneously into a free proton and a free electron. The mass difference (4.76) is called the *binding energy* of the hydrogen atom.

Actually the capture of the free electrons into atomic orbits and the formation of hydrogen occurs only at 0.3 eV, because the released binding energy reheats the remaining electrons, and also because the large amount of entropy in the Universe favours free protons and electrons. When this *recombination* process is completed, the photons find no more free electrons to scatter against, and they do not scatter against the neutral hydrogen atoms. Thus the photons decouple, continuing in straight lines from the point of their last scattering with a mean free path exceeding the Hubble radius. Hence the Universe becomes transparent to radiation.

In this context the term *re*combination is slightly misleading, because the electrons have never been combined into atoms before. The term comes from laboratory physics where free electrons and *ionized* atoms are created by heating matter. Subsequently when the matter cools, the electrons and ions recombine into atoms.

The last scattering of the photons occurred while the Universe cooled from 5000 K to 3000 K. The location today of this *last scattering surface* is at an average radial distance corresponding to a redshift of $z \simeq 1100$. The last scattering did not occur to all photons at the same time, so this surface is really a shell of thickness $\Delta z \simeq 0.07\, z$. The time of last scattering may be taken to be 180 000 $(\Omega_0 h^2)^{-1/2}$ years corresponding to a photon energy of about 0.26 eV.

Thus Big Bang cosmology makes some very important predictions: the Universe today should still be filled with primordial photon and neutrino radiation, and both should have a blackbody spectrum (4.3) with a temperature related to the age of the Universe. This radiation should be essentially isotropic since it originated in the now spherical last scattering surface. In particular, it should be uncorrelated to the radiation from foreground sources of later date, such as our Galaxy. That these predictions are verified for the photons (but not yet for the neutrinos) we shall see in Chapter 5.

Problems

1. Show that an expansion by a factor S leaves the blackbody spectrum (4.3) unchanged, except that T decreases to T/S.
2. What is the ratio of the electric force to the gravitational force between two protons? [2]
3. Use the definition of entropy in Eq. (4.7) and the law of conservation of energy Eq. (3.33) to show what functional forms of equations of state lead to conservation of entropy.
4. The flow of total energy received on Earth from the Sun is expressed by the *solar constant* 1.36×10^6 erg/cm^{2}s. Use Eq. (4.43) to determine the surface temperature of the Sun. Determine *Wien's constant*

$$b = \lambda T \tag{4.77}$$

   using this temperature and the knowledge that the dominant colour of the Sun is yellow with a wavelength of $\lambda = 0.503$ μm. What energy density does that flow correspond to ?
5. The random velocity of galaxies is roughly 100 km/s, and their number density is 0.0029 per cubic light year. If the average mass of a galaxy is 3×10^{44}g, what is the pressure of a gas of galaxies ? What is the temperature? [3]
6. A line in the spectrum of hydrogen has frequency $\nu = 2.5 \times 10^{15}$ Hz. If this radiation is emitted by hydrogen on the surface of a star where the temperature is 6000 K, what is the Doppler broadening? [3]
7. A spherical satellite of radius r painted black, travels around the Sun at a distance d from the centre. The Sun radiates as a blackbody at a temperature of 6000 K. If the Sun subtends an angle of θ radians as seen from the satellite (with $\theta \ll 1$), find an expression for the equilibrium temperature of the satellite in terms of θ. To proceed, calculate the energy absorbed by the satellite, and the energy radiated per unit time [3].
8. Use the laws of conservation of energy and momentum and the equation of relativistic kinematics (2.54) to show that positronium cannot decay into a single photon. What is the mass and the momentum of the virtual photon in reaction (4.34)?
9. Use the equation of relativistic kinematics (2.54) to calculate the energy and velocity of the muon from the decay (4.56) of a pion at rest. The neutrino is assumed to be massless.
10. The capture cross section σ for neutrinos is about 10^{-44} cm^2 per nucleon. Calculate the mean free path $\ell = 1/\sigma\rho$ for neutrino capture at the centre of the Sun, where $\rho \approx 100$ g/cm^3. Convert your answer to parsecs and compare with the radius of the Sun [2].
11. What are the possible decay modes of the τ^- lepton ?
12. Calculate the energy density represented by the rest mass of all the electrons in the Universe at the time of photon reheating when the kinetic energy of electrons is 0.2 MeV.
13. When the pions disappear below 140 MeV because of annihilation and decay, some reheating of the remaining particles occurs due to entropy conservation. Calculate the temperature increase factor.

References

1. E. W. Kolb and M. S. Turner, *The Early Universe*, Addison-Wesley Publ. Co., Reading, Mass, 1990.
2. F. H. Shu, *The Physical Universe*, University Science Books, Mill Valley, CA, 1982.
3. S. Gasiorowicz, *The Structure of Matter*, Addison-Wesley Publ. Co., Reading, Mass, 1979.

5 Relics of the Big Bang

So far we have mainly worked on theory, establishing qualitatively the standard model of Big Bang. The inputs from astronomical observations have been quite general, such as the homogeneity and isotropy of matter embodied in the cosmological principle, and the observation that stellar bodies exert gravitational attraction on each other. A notable observational input was Hubble's law which led to important quantitative conclusions such as the age of the Universe. Further experimental input has come from other fields of science, such as particle physics and thermodynamics.

It is necessary to understand theory in order to grasp the implications of observational results, although the chronology of research does not always follow that order. In Section 5.1 we shall meet the important discovery of the *cosmic microwave background* (CMB) radiation by Penzias and Wilson which they themselves did not understand because they were unaware of the theory of Big Bang which in fact predicted what they oberved. The most important determination of the temperature and distribution of this relic blackbody radiation comes today from a 1990 satellite experiment, the Cosmic Background Explorer, or COBE.

Since the discovery of CMB in 1964, numerous experiments have been carried out trying to reveal relic anisotropies in the CMB at large angular scales, i.e. in directions differing by 2° or more. There are good theoretical reasons to expect such anisotropies. Except for the dipole anisotropy which is due to Earth's motion relative to the CMB, all experiments have only confirmed the absence of anisotropies down to temperature variations as small as 50 μK. In Section 5.2 we tell the story of the discovery in 1992 of the anisotropies and the present situation at different angular scales.

In Section 5.3 we return to the thermal history of the Universe for the momentous process of nucleosynthesis which has left us very important clues in the form of relic helium and other light nuclei. The nucleosynthesis is really a very narrow bottleneck for all cosmological models, and one which has amply confirmed the standard Big Bang model. We find that baryonic matter present since nucleosynthesis is completely insufficient to close the Universe.

5.1 The Cosmic Microwave Background

In 1948 Georg Gamow, Ralph Alpher and Robert Herman undertook to calculate the present temperature of the primordial blackbody radiation, using essentially the formalism we met in the previous Chapter. They found that the CMB should still exist today, but cooled in the process of expansion to the very low temperature of $T_0 \approx 5$ K. This corresponds to a photon wavelength of

$$\lambda = \frac{hc}{kT_0} = 2.9 \text{ mm}. \tag{5.1}$$

This is in the microwave range of radio waves (see Table 3 on page 71).

We can now redo their calculation, using some hindsight. Let us first recall from Eqs. (3.47) and (3.48) that the expansion rate changed at the moment when radiation and matter contributed equally to the energy density. For our calculation we need to know this equality time, t_{eq}, and the temperature T_{eq}. The energy density in radiation is given by Eq. (4.46),

$$\varepsilon_r = (g_\gamma + 3g_\nu)\frac{a}{2} \, T_{eq}^4 \, . \tag{5.2}$$

The energy density of matter at time T_{eq} is given by Eq. (4.25), except that the electron (e^- and e^+) energy needs to be averaged over the spectrum (4.41). We could in principle solve for T_{eq} by equating radiation and matter densities,

$$\varepsilon_r(T_{eq}) = \varepsilon_m(T_{eq}) \, . \tag{5.3}$$

However, the number density n of matter is a quantity to be discussed later when we return to the fate of nucleons, so it would be premature to consider its value to be known here. Also, equation (5.3) is a rather complicated and non-linear function of T_{eq}, so we shall not make the effort to solve it.

The transition epoch happens to be close to the recombination or last scattering. Let us take t_{eq} to be approximately at 350 000 yr and $T_{eq} = 3000$ K for the time being. Actually the exact moment of transition depends on the poorly known value of Ω_0, as one can see from Eq. (4.15). If we take the present age of the Universe to be $t_0 = 15$ Gyr, Eq. (3.47) gives the present temperature of the CMB,

$$T_0 = T_{eq} \left(\frac{t_{eq}}{t_0} \right)^{2/3} = 2.45 \text{ K}. \tag{5.4}$$

This is very close to the observed value, as we shall see.

Nobody paid much attention to the prediction of Gamow, Alpher and Herman, because the Big Bang theory was generally considered to be wildly speculative, and detection of the predicted radiation was far beyond the technical capabilities existing at that time. In particular, their prediction was not known in 1964 to Arno Penzias and Robert Wilson who were testing a sensitive antenna intended for satellite communication. They wanted to calibrate it in an environment free of all radiation, so they chose a wavelength of $\lambda = 7.35$ cm in the relatively quiet window between the known emission from the

Galaxy at longer wavelengths and the emission at shorter wavelengths from the Earth's atmosphere. They also directed the antenna high above the galactic plane where scattered radiation from the Galaxy would be minimal.

To their consternation and annoyance they found a constant low level of background noise in every direction. This radiation did not seem to be of galactic origin, because then they should have seen an intensity peak in the direction of the nearby M31 galaxy in the Andromeda. It could also not have originated in Earth's atmosphere, because such an effect would vary with the altitude above the horizon as a function of the thickness of the atmosphere.

Thus Penzias and Wilson suspected technical problems with the antenna (in which a couple of pigeons turned out to be roosting) or with the electronics. All searches failing they finally concluded, correctly, that the Universe was uniformly filled with an 'excess' radiation corresponding to a blackbody temperature of 3.5 K, and that this radiation was isotropic and unpolarized within their measurement precision.

At Princeton University a group of physicists led by Robert Dicke had at that time rediscovered independently the theory of Gamow and collaborators, and they were preparing to measure the CMB radiation when they heard of the remarkable 3.5 K 'excess' radiation. The results of Penzias' and Wilson's measurements were subsequently published in 1965 jointly with an article by Dicke and collaborators which explained the cosmological implications. The full story is told by Peebles [1] who was a student of Dicke at that time. Penzias and Wilson (but not Gamow or Dicke) were subsequently awarded the Nobel prize in 1978 for this discovery.

This evidence for the 15 Gyr old echo of the Big Bang counts as the most important discovery in cosmology since Hubble's law. In contrast to all radiation from astronomical bodies which is generally hotter, and which has been emitted much later, the CMB has existed since the era of radiation domination. It is hard to understand how the CMB could have arisen without the cosmic material having once been highly compressed and exceedingly hot. At all times after recombination there is no known mechanism which could have produced a blackbody spectrum in the microwave range, because the Universe is transparent to radio waves.

In principle, one intensity measurement at an arbitrary wavelength of the blackbody spectrum (4.3) is sufficient to determine its average temperature T, because this is the only free parameter. On the other hand, one needs measurements at different wavelengths to establish that the spectrum is indeed blackbody.

Although several accurate experiments since Penzias and Wilson have confirmed the temperature near 3 K by measurements at different wavelengths, the conclusive demonstration that the spectrum is indeed blackbody also in the region masked by radiation from the Earth's atmosphere was made by a dedicated instrument, the Far Infrared Absolute Spectrophotometer (FIRAS) aboard the Cosmic Background Explorer satellite (COBE) launched in 1989 [2].

The temperature deduced from the COBE measurements [3] is

$$T_0 = 2.726 \text{ K} \tag{5.5}$$

with an accuracy of 0.01 K. All theories that attempt to explain the origin of large-scale structure seen in the Universe today must conform to the constraints imposed by these COBE measurements.

The spectrum reported by the COBE team in 1993 [3] shown in Fig. 26 matches the predictions of the hot Big Bang theory to an extraordinary degree. The measurement errors on each of the 34 wavelength positions are so small that they cannot be distinguished from the theoretical blackbody curve. It is worth noting that such a pure blackbody spectrum had never been observed in laboratory experiments.

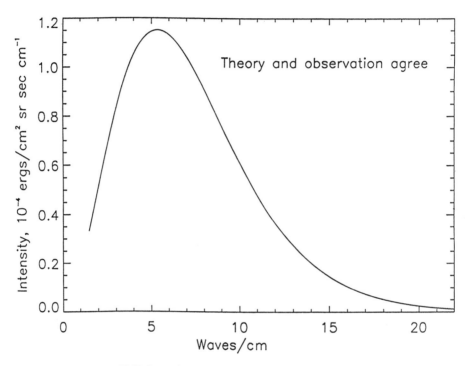

Fig. 26 Spectrum of the CMB from data taken with the Far Infrared Absolute Spectrophotometer (FIRAS) instrument aboard NASA's Cosmic Background Explorer (COBE). Reproduced from reference [2] by permission of the COBE Science Working Group

Given this precise value of T_0, one can determine several important quantities. From Eq. (4.46) one can calculate the present energy density in radiation

$$\varepsilon_{r,0} = \frac{g_{eff}}{2} \, a T_0^4 \simeq 1.40 \times 10^{-12} c^{-2} \text{ erg cm}^3 \simeq 7.80 \times 10^{-34} \text{ g cm}^3 \, . \qquad (5.6)$$

For g_{eff} we have not used the value obtained from Eq. (4.45), but a corrected value

$$g_{eff}(T_0) = 2 + 3 \times \frac{7}{4} \left(\frac{4}{11} \right)^{4/3} = 3.36 \qquad (5.7)$$

which takes into account that the neutrinos have a lower temperature than the photons by the factor $(4/11)^{1/3}$, and which assumes that the mass of the neutrinos is less than T_0.

The present value of the density parameter for radiation is then

$$\Omega_{r,0} = \frac{\varepsilon_{r,0}}{\rho_c c^2} \simeq 4.11 \times 10^{-5} h^{-2} . \tag{5.8}$$

Obviously the radiation energy is very small today and far from the value $\Omega = 1$ required to close the Universe. From this value and Eq. (4.15) we can determine the time of equality of radiation and matter density assuming $t_0 = 15$ Gyr,

$$t_{eq} = \left(\frac{\Omega_{r,0}}{\Omega_0} \frac{3.36}{3.36 + 3.5} \right)^{\frac{3}{2}} t_0 \simeq 4.26 \times 10^{10} \ (\Omega_0 h^2)^{-\frac{3}{2}} \text{ s.} \tag{5.9}$$

The entropy density is given by Eq. (4.67). Now, however, the temperature dependence of g_{eff} is a power 3 rather than a power 4 as in Eq. (5.7), so the factor $(4/11)^{4/3}$ becomes just $4/11$, and g_{eff} becomes 3.91 (see e.g. reference [4]). This corresponds to the curve g_{*S} in Fig. 24. Thus the present value of the entropy density is

$$\frac{s}{V} = \frac{4}{3} \frac{\varepsilon_{r,0}}{kT_0} \frac{3.91}{3.36} \simeq 2888 \text{ cm}^{-3} . \tag{5.10}$$

From this and Eq. (4.74) we can find the present number density of CMB photons,

$$N_\gamma = \frac{s/V}{1.80 \times 3.91} = 410 \text{ photons cm}^3 . \tag{5.11}$$

Using the known temperature dependences of the above quantities one can work backwards to find their values at earlier times. In Fig. 19 we plot the energy density for matter and radiation as a function of the scale S of the Universe, from about one second after Big Bang until now. The above quantities and other derived present properties of the Universe are collected in Table 6 on page 112.

It also follows from the above value of the temperature and the relation (4.73) that the neutrino temperature today is $T_\nu = 1.946$ K. Unfortunately nobody has yet been able to devise a method to observe the relic neutrino background.

5.2 Anisotropies in the Background Radiation

The temperature measurement of Penzias' and Wilson's antenna was not very precise by today's standards. Their conclusion about the isotropy of the CMB was based on an accuracy of only 1.0 K. When the measurements improved over the years it was found that the CMB exhibited a *dipole anisotropy*. The temperature varies minutely over the sky in such a way that it is maximally blueshifted in one direction (call it α) and maximally redshifted in the opposite direction ($\alpha + 180°$). In a direction θ it is

$$T(\theta) = T(\alpha)(1 + \nu \cos \theta) , \tag{5.12}$$

where ν is the amplitude of the dipole anisotropy. Although this shift is small, only $\nu T(\alpha) \approx 3.3$ mK, it is very accurately measured.

At the end of the previous chapter we concluded that the CMB should be essentially isotropic since it originated in the last scattering surface which has now receded to a distance of $z \simeq 1100$ in all directions. Note that the most distant quasars have red shifts of the order of $z = 4 - 5$. Their distance to the last scattering surface is actually much closer than their distance to us: $z = 4$ corresponds to about 90% of the distance to the last scattering surface.

In the standard model the expansion is spherically symmetric, so it is quite clear that the dipole anisotropy cannot be of cosmological origin. Rather, it is well explained by a Doppler shift: we are travelling 'against' the radiation in the direction of maximal blueshift with relative velocity ν.

Thus there is a frame in which the CMB is isotropic—not a rest frame since radiation cannot be at rest. This frame is then comoving with the expansion of the Universe. We already referred to it in Section 2.2 where we noted that to a fundamental observer at rest in the comoving frame the Universe must appear isotropic if it is homogeneous. Although general relativity was constructed to be explicitly frame-independent, the comoving frame in which the CMB is isotropic is observationally convenient. The fundamental observer is at position B in Fig. 27.

The interpretation today is that not only does the Earth move around the sun, and the solar system participates in the rotation of the Galaxy, but also the Galaxy moves relative to our Local Galaxy Group which in turn is falling towards a centre behind the Hydra-Centaurus supercluster in the constellation Virgo. From the observation that our motion relative to the CMB is about 365 km/s, these velocity vectors add up to a peculiar motion of the Galaxy of about 550 km/s, and a peculiar motion of the Local Group of about 630 km/s [10]. Thus the dipole anisotropy seen by the Earth-based observer A in Fig. 27 tells us that we and the Local Group are part of a larger, gravitationally bound system.

In the previous chapter we argued that thermal equilibrium could be established throughout the Universe during the radiation era because photons could traverse the whole Universe and interactions could take place in a time much shorter than a Hubble time. However, there is a snag to this argument: the conditions at any spacetime point can only be influenced by events within its past light cone.

Since the time of last scattering the spatial width of the past light cone and the particle horizon have grown in linear proportion to the longer time perspective. At the same time all linear distances, such as the particle horizon at the time of last scattering have grown with the expansion in proportion to the $\frac{2}{3}$-power of time (actually this power law has been valid since t_{eq}, but t_{LSS} and t_{eq} are nearly simultaneous). The net effect is that the particle horizon we see today covers regions which were causally disconnected at earlier times.

At the time of last scattering the Universe was 1100 times smaller than now, and the time perspective back to the Big Bang was only the fraction $t_{LSS}/t_0 \simeq 2.3 \times 10^{-5}$ of our perspective. The particle horizon at the last scattering surface, σ_p, is obtained by substituting $S(t) \propto (t_{LSS}/t)^{1/2}$ into Eq. (2.46) and integrating from zero time to t_{LSS},

$$\sigma_p \propto \int_0^{t_{LSS}} dt \left(\frac{t_{LSS}}{t} \right)^{1/2} = 2t_{LSS}. \tag{5.13}$$

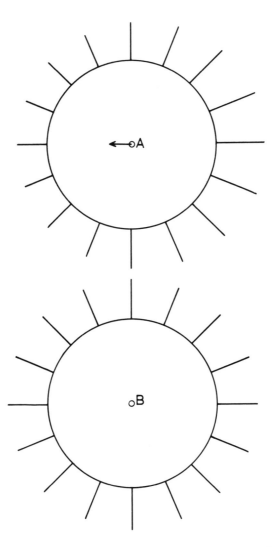

Fig. 27 The observer A in the solar rest frame sees the CMB to have dipole anisotropy—the length of the radial lines illustrate the CMB intensity—because he is moving in the direction of the arrow. The fundamental observer at position B has removed the anisotropy

It is not very critical what we call 'zero' time, the lower limit of the integrand has essentially no effect even if it is chosen as late as $10^{-4}t_{LSS}$.

The size of the *event horizon* at the time of last scattering, σ_e, is obtained by substituting $S(t) \propto (t_{LSS}/t)^{2/3}$ into Eq. (2.9) and integrating over the full epoch of matter-domination from t_{LSS} to $t_{max} = t_u$. Assuming flat space $k = 0$, we have

$$\sigma_e \propto \int_{t_{LSS}}^{t_u} dt \left(\frac{t_{LSS}}{t}\right)^{1/2} = 3t_{LSS}\left[\left(\frac{t_u}{t_{LSS}}\right)^{1/3} - 1\right]. \qquad (5.14)$$

Let us take $t_{LSS} = 0.35$ Myr and $t_u = 15$ Gyr. Then the LSS particle horizon σ_p is seen today as an arc on the periphery of our particle horizon, subtending an angle

$$\frac{180}{\pi} \left[\frac{\sigma_p}{\sigma_e} \right]_{LSS} \simeq 1.12° \,. \tag{5.15}$$

This is illustrated in Fig. 28 which, for clarity, is not drawn to scale.

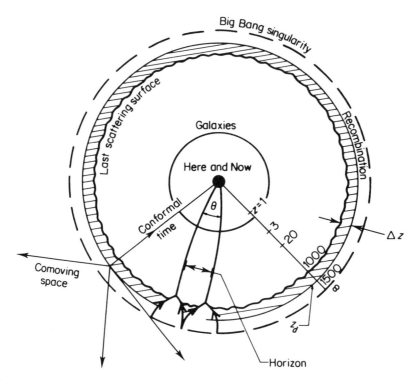

Fig. 28 A co-moving space/conformal time diagram of the Big Bang. The observer (here and now) is at the centre. The Big Bang singularity has receded to the outermost dashed circle, and the horizon scale is schematically indicated at last scattering. It corresponds to an arc of angle θ today. Reproduced from reference [5] by permission of J. Silk

It follows that the temperature of the CMB radiation coming from any $1°$ arc could not have been causally connected to the temperature on a neighbouring arc, so there is no reason why they should be equal. Thus one would expect to observe variations in the CMB on scales of $1°$ or more, indicating that these early mini-universes had different temperatures at the time of emission of the photons.

A second reason for expecting anisotropy is the fact that matter today is not very homogeneously distributed. There are galaxies, groups of galaxies, supergalaxies and strings of supergalaxies with great voids in between. At the time of last scattering some lumpiness in the energy density must have been the 'seeds' or *progenitors* of these cosmic structures, and one would expect to see traces of that lumpiness in the CMB. The angular scale subtended by progenitors corresponding to the largest cosmic structures known, of size perhaps 200 h^{-1} Mpc, is of the order of 3°. Note that the CMB photons from the progenitors of galaxies were emitted long before the galaxies started to radiate light.

We noted before that the last scattering did not occur to all photons at exactly the same time. There may have been adiabatic fluctuations in the mass density causing temperature variations, visible today as anisotropies on scales of $\lesssim$30′. The LSS may also have been in motion, causing Doppler shifts manifested today as anisotropies on scales of 30′ to 2°.

Photons 'climbing out' of the gravitational potential well caused by a lump of higher density are redshifted by an amount given by Eq. (2.65). Inversely, photons emitted from regions of low density 'roll down' from the gravitational potential, and are blueshifted. In the long passage to us they may traverse further regions of gravitational fluctuations, but then their frequency shift upon entering the potential is compensated for by an opposite frequency shift when leaving it. Thus the CMB photons preserve a 'memory' of the density fluctuations at emission, manifested today as temperature variations at large angular scales. A possible disturbance on angular scales less than a few degrees are long wavelength gravitational waves [11].

Thus there are many contributing mechanisms, but the dominating effect on large scales is the *Sachs–Wolfe effect* which makes $\delta T/T$ linearly dependent on the density fluctuations $\delta\rho/\rho$,

$$\frac{\delta T}{T} \simeq \frac{1}{3}\left(\frac{L_{dec}}{ct_{dec}}\right)^2 \frac{\delta\rho}{\rho} \,, \qquad (5.16)$$

where L_{dec} is the size of the structure at decoupling time t_{dec}.

Let us now turn to the observational evidence. For many years microwave experiments have been trying to detect temperature variations on angular scales ranging from a few arcminutes to tens of degrees. Ever increasing sensitivities had brought down the limits on $\delta T/T$ to near 10^{-5} without finding any evidence for anisotropy until the year 1992. At that time the first results of large-scale anisotropy was reported by the COBE space experiment [6].

On board the COBE satellite there are several instruments of which one, the Differential Microwave Radiometer, has two antennas with 7° opening angles directed 60° apart. Comparing the signals from the two antennas, this instrument is sensitive to anisotropies on very large angular scales. This instrument has been mapping the CMB at three frequencies, covering the full sky every six months. To avoid unwanted radiation the antennas point at least 20° above the galactic plane and away from the Sun. In addition, the measurements have to be corrected for a host of minor sources of errors such as instrumental imbalance and microwave emission from the Moon and Jupiter and from the Galaxy at high latitudes. Also the Earth's rotation and magnetic field need to be taken into account as well as the dipole anisotropy due to the peculiar motion of the galaxy 'against' the CMB radiation field.

Its first published results [6] cover one year of measurements or two complete mappings of the full sky followed by the above spherical harmonic analysis. For the first time ever anisotropies were found, corresponding to temperature variations of

$$\delta T = 30 \pm 5 \ \mu\text{K, or} \ \ \delta T/T = 1.1 \times 10^{-5} \,. \qquad (5.17)$$

Since the precision of the COBE measurements surpasses all previous experiments one can well understand that such small temperature variations had not been seen before. The importance of this discovery was succinctly emphasized by the COBE team who wrote that 'a new branch of astronomy has commenced'.

Half of the above $\delta T/T$ value, or 0.5×10^{-5}, could be ascribed to *quadrupole anisotropy* at $90°$ angular scale. Although some quadrupole anisotropy is kinetic, related to the dipole anisotropy and the motion of Earth, this term could be subtracted. The remainder is then quadrupole anisotropy of purely cosmological origin which could have arisen if the expansion had not been spherically symmetric. This would clearly contradict the cosmological principle, and of course also the standard Friedmann–Robertson–Walker cosmology which is based on it. Anisotropies on scales of $90°$ could also not have been caused by density fluctuations at the time of last scattering because, as we have seen, no causal processes can exceed the particle horizon, or $2°$.

The other half of the above $\delta T/T$ value, or 0.6×10^{-5}, are intrinsic CMB fluctuations on all scales, indicating the existence of structures of 500 Mpc size. This is much larger than any optically observed supercluster.

Fluctuations on such scales could only have been generated today inside our horizon of size σ_0. Thus we have an indication that the standard model must be modified at early times, somehow enlarging the horizon to fit large-scale density fluctuations. This leads to models of *cosmic inflation* to which we shall come in Chapter 8.

Temperature fluctuations around a mean temperature in a direction α on the sky can be analyzed in terms of the *autocorrelation function* $C(\theta)$ which measures the average product of temperatures in two directions separated by an angle θ,

$$C(\theta) = \langle T(\alpha)T(\alpha + \theta) \rangle \ . \tag{5.18}$$

The temperature autocorrelation function can be expressed as a sum of *Legendre polynomials* $P_\ell(\theta)$ of order ℓ with coefficients or *powers* a_ℓ^2,

$$C(\theta) = \frac{1}{4\pi} \sum_{\ell=2}^{\infty} a_\ell^2 (2\ell + 1) P_\ell(\theta) \ . \tag{5.19}$$

The COBE analysis starts with $\ell \geq 2$ because one subtracts the $\ell = 1$ mode which is the dipole.

In the analysis the powers a_ℓ^2 are adjusted to give a best fit of $C(\theta)$ to the observed temperature. The resulting distribution of a_ℓ^2 values versus ℓ is called the *power spectrum* of the fluctuations. The higher the angular resolution, the more terms of high ℓ must be included. The Differential Microwave Radiometer in COBE is sensitive to the low power spectrum with Legendre coefficients up to about 20.

However, a different story is told by the very preliminary observations on the South Pole [7]. This experiment is sensitive to the high end of the power spectrum, with Legendre coefficients from about 40 to 400, corresponding to an angular range of $1° - 2°$. Thus it covers a completely different angular and spectral range than COBE, so it can neither confirm nor disprove the COBE result. At the time of writing it has seen no anisotropies down to an amplitude of

$$\delta T/T \stackrel{<}{\sim} 7 \times 10^{-6} \ . \tag{5.20}$$

In Fig. 29 we show the results of the COBE and the South Pole experiments, as well as upper limits quoted by a team at MIT [8] and by an experiment from the Observatorio del Teide in Tenerife [9] at 5.6°. This latter experiment which is centred on Legendre coefficients between about 8 and 35 reports the limit

$$\delta T/T \stackrel{<}{\sim} 1.8 \times 10^{-5} \ . \tag{5.21}$$

These four experiments do not really conflict with each other because each of them depends on statistical and systematic errors which, when taken into account, do allow a common likely value near 1.0×10^{-5}. We shall return to a discussion of these results in Chapter 9.

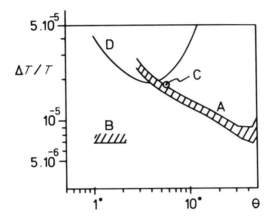

Fig. 29 Some recent measurements of and searches for the CMB anisotropy $\Delta T/T$ as a function of correlation angle θ. The results A,B,C,D are from references [5], [6], [8], [7], respectively

5.3 Nucleosynthesis

We left the story of the decoupling nucleons in Section 4.5 at the time when the weak interactions (4.53) and (4.54) ceased and the conversion of protons to neutrons stopped because the energy in the thermal bath dropped below 0.8 MeV. The neutrons and protons were then non-relativistic, so their number densities were each given by Maxwell–Boltzmann distributions (4.42). Their ratio in equilibrium is given by

$$\frac{N_n}{N_p} = \left(\frac{m_n}{m_p}\right)^{3/2} \exp\left(-\frac{m_n - m_p}{kT}\right) \ . \tag{5.22}$$

At energies of the order of $m_n - m_p = 1.293$ MeV or less, this ratio is dominated by the exponential. Thus at $kT = 0.8$ MeV one finds that the ratio has dropped from 1 to 1/5.

Already at a few MeV, nuclear *fusion reactions* start to build up light elements. These reactions are exothermic: when a neutron and a proton fuse into a bound state some of the nucleonic matter is converted into pure energy according to Einstein's formula (2.55). This binding energy of the *deuteron d*

$$m_p + m_n - m_d = 2.22 \text{ MeV}$$

is liberated in the form of radiation,

$$n + p \rightarrow d + \gamma \ . \tag{5.23}$$

The deuteron is also written $^2\text{H}^+$ in general nuclear physics notation where the superscript $A = 2$ indicates the number of nucleons, and the electric charge is given by the superscript $+$. The bound state formed by a deuteron and an electron is the *deuterium* atom ^2H which of course is electrically neutral. Although the deuterons are formed in very small quantities, they are of crucial importance to the final composition of matter.

As long as photons of 2.22 MeV or more are available, the reaction (5.23) can go the other way: the deuterons *photodisintegrate* into free protons and neutrons. Even when the mean temperature of radiation drops considerably below 2.22 MeV, there is still a high-energy tail of the Planck distribution containing hard γ-rays which destroy the deuterons as fast as they are produced.

All evidence suggests that the number density of baryons, or equivalently nucleons, is today very small. In particular, we are able to calculate it to a factor $\Omega_B h^2$,

$$N_B = \frac{\rho_B}{m_B} = \frac{\Omega_B \rho_c}{m_B} \simeq 1.13 \times 10^{-5} \ \Omega_B h^2 \ \text{cm}^{-3} \ . \tag{5.24}$$

The baryon density parameter Ω_B cannot be very much larger than unity because that would close the Universe too fast, and h is less than unity, so the numerical factor is close to an upper limit to N_B.

In Eq. (5.11) we deduced that the photon number density today is 410 per cm^3. Thus the ratio of baryons to photons is

$$\eta \equiv \frac{N_B}{N_\gamma} \simeq 2.76 \times 10^{-8} \ \Omega_B h^2 \ . \tag{5.25}$$

It is clear from this very small figure that an extremely tiny fraction of the high-energy tail of the photon distribution may contain enough hard γ-rays to photodisintegrate the deuterons.

An even more serious obstacle to permanent deuteron production is the high entropy per nucleon in the Universe. Each time a deuteron is produced the degrees of freedom decrease, and so the entropy must be shared among the remaining nucleons. This raises their temperature, counteracting the formation of deuterons. Detailed calculations show that deuteron production becomes thermodynamically favoured only at 0.07 MeV. Thus, although deuterons are favoured on energetic grounds already at 2 MeV, free nucleons continue to be favoured by the high entropy down to 0.07 MeV.

Other nuclear fusion reactions also commence at a few MeV. The *npp* bound state $^3\text{He}^{++}$ is produced in the fusion of two deuterons,

$$d + d \rightarrow {}^3\text{He}^{++} + n \,, \tag{5.26}$$

$$p + d \rightarrow {}^3\text{He}^{++} + \gamma \,, \tag{5.27}$$

where the final-state particles share the binding energy $2m_p + m_n - m({}^3\text{He}^{++}) = 7.72$ MeV. This reaction is also hampered by the large entropy per nucleon, so it becomes thermodynamically favoured only at 0.11 MeV.

The *nnp* bound state ${}^3\text{H}^+$, or *triton t*, is the ionized *tritium* atom, ${}^3\text{H}$. It is produced in the fusion reactions

$$n + d \rightarrow t + \gamma \,, \tag{5.28}$$

$$d + d \rightarrow t + p \,, \tag{5.29}$$

$$n + {}^3\text{He} \rightarrow t + p \,, \tag{5.30}$$

with the binding energy $m_p + 2m_n - m_t = 8.48$ MeV.

A very stable nucleus is the *nnpp* bound state ${}^4\text{He}^{++}$ with a very large binding energy, $2m_p + 2m_n - m({}^4\text{He}^{++}) = 28.3$ MeV. Once its production is favoured by the entropy law, at about 0.28 MeV, there are no more γ-rays left hard enough to photodisintegrate it. From the examples set by the deuteron fusion reactions above, it may seem that ${}^4\text{He}^{++}$ would be most naturally produced in the reaction

$$d + d \rightarrow {}^4\text{He}^{++} + \gamma \,. \tag{5.31}$$

However, ${}^3\text{He}^{++}$ and ${}^3\text{H}^+$ production is preferred over deuteron fusion, so ${}^4\text{He}^{++}$ is only produced in a second step when these nuclei become abundant. The reactions are then

$$n + {}^3\text{He}^{++} \rightarrow {}^4\text{He}^{++} + \gamma \,, \tag{5.32}$$

$$d + {}^3\text{He}^{++} \rightarrow {}^4\text{He}^{++} + p \,, \tag{5.33}$$

$$p + t \rightarrow {}^4\text{He}^{++} + \gamma \,, \tag{5.34}$$

$$d + t \rightarrow {}^4\text{He}^{++} + n \,. \tag{5.35}$$

The delay before these reactions start is often referred to as the *deuterium bottleneck*.

Below 0.8 MeV occasional weak interactions in the high-energy tails of the lepton and nucleon Fermi distributions reduce the n/p ratio further, but no longer by the exponential factor in Eq. (5.22). At 0.1 MeV when the temperature is 1.2×10^9 K and the time elapsed since Big Bang is a little over 2 minutes, the beta decays of neutrons already noticeably convert neutrons to protons. The mean life of free neutrons is 14.8 minutes. At this point the n/p ratio has reached its final value

$$\frac{N_n}{N_p} \simeq \frac{1}{7} \,. \tag{5.36}$$

In Fig. 30 the temperature dependence of this ratio is shown, as well as the equilibrium (Maxwell–Boltzmann) ratio.

These remaining neutrons have no time to decay before they fuse into deuterons and subsequently into ${}^4\text{He}^{++}$. There they stay until today because bound neutrons do not decay. The same number of protons as neutrons go into ${}^4\text{He}$, and the remaining free protons are

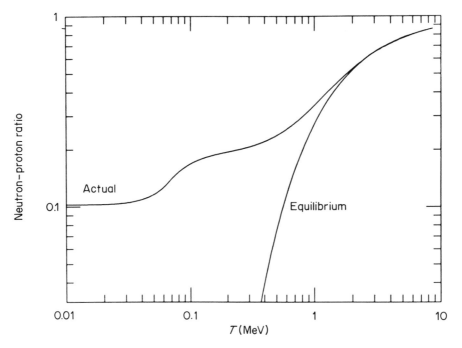

Fig. 30   The equilibrium and actual values of the *n/p* ratio. Reproduced from reference [3] by permission of E. W. Kolb and M. S. Turner

the nuclei of future hydrogen atoms. Thus the end result of the nucleosynthesis taking place between 100 and 700 seconds after Big Bang is a Universe composed almost entirely of hydrogen and helium ions. Why not heavier nuclei?

It is an odd circumstance of Nature that although there exist stable nuclei composed of $A = 1, 2, 3$ and 4 nucleons, no nucleus of $A = 5$ exists, and no stable one with $A = 8$. In between these gaps, there exist the unstable nuclei ^{6}Li and ^{7}Be, and the stable ^{7}Li. Because of these gaps and because ^{4}He is so strongly bound, nucleosynthesis essentially stops after ^{4}He production. Only minute quantities of the stable nuclei ^{2}H, ^{3}He, and ^{7}Li can be produced.

The fusion rate at energy E of two nuclei of charges Z_1, Z_2 is proportional to the *Gamow penetration factor*

$$\exp\left(-\frac{2Z_1Z_2}{\sqrt{E}}\right) .$$ (5.37)

Thus as the energy decreases, the fusion of nuclei other than the very lightest ones becomes rapidly improbable.

The relic abundances of the light elements bear a quite important testimony of the n/p ratio at the time of the nucleosynthesis. In fact, this is the earliest testimony of the Big Bang we have. Only 350 000 years later recombination occurred when all the stable ions produced captured electrons to become neutral atoms. The CMB testimony of the conditions at the last scattering surface is from that time.

From the ratio (5.36) we obtain immediately the ratio of ^{4}He to ^{1}H,

$$X_4 \equiv \frac{N(^4\text{He})}{N(^1\text{H})} = \frac{N_n/2}{N_p - N_n} \simeq \frac{1}{12} \ . \tag{5.38}$$

The number of ^{4}He nuclei is clearly half the number of neutrons when the minute amounts of ^{2}H, ^{3}He and ^{7}Li are neglected. The same number of protons as neutrons go into ^{4}He thus the excess number of protons becoming hydrogen is $N_p - N_n$. Usually one quotes the ratio of mass in ^{4}He to total mass in ^{1}H and ^{4}He,

$$Y_4 \equiv \frac{4X_4}{1 + 4X_4} \simeq 0.25 \ . \tag{5.39}$$

A present best estimate of Y_4 from observational data is in the range 0.22–0.24 with no directional variation. This is a strong support of the Big Bang hypothesis.

The helium mass abundance Y_4 depends sensitively on several parameters. The quantity η in Eq. (5.25), which is the ratio of baryon and photon number densities, is proportional to the baryon density parameter Ω_B. If the number of baryons increases, Ω_B and η also increase, and the entropy per baryon decreases. Remembering that the large entropy per baryon was the main obstacle to early deuteron and helium production, the consequence is that helium production can start earlier. But then the neutrons would have had less time to β-decay, so the n/p ratio would be larger than 1/7. It follows that more helium will be produced: Y_4 increases.

The abundances of the other light elements depend similarly on η. They also depend on the neutron mean life τ_n and on the number of neutrino families F_ν, both of which were poorly known until 1990. Although τ_n is now known to better than a quarter percent, and F_ν is known to be 3 to within 0.9 percent, it may be instructive to follow the arguments about how they affect the value of Y_4.

Let us rewrite the decoupling condition Eq. (4.64) for neutrons

$$\frac{\Gamma_{wi}}{H} = A T_d^3 \ , \tag{5.40}$$

where A is the proportionality constant left out in Eq. (4.64) and T_d is the decoupling temperature. An increase in the neutron mean life implies a decrease in the reaction rate Γ_{wi} and therefore a decrease in A. At temperature T_d the ratio of the reaction rate to the expansion rate is unity, thus

$$T_d = A^{-1/3} \ . \tag{5.41}$$

Hence a longer neutron mean life implies a higher decoupling temperature and an earlier decoupling time. As we have already seen, an earlier start of helium production leads to an increase in Y_4.

The expansion rate H of the Universe is, according to Eqs. (4.48) and (4.50), proportional to $\sqrt{g_{\textit{eff}}}$, which in turn depends on the number of neutrino families F_ν. In Eqs. (4.51) and (5.7) we had set $F_\nu = 3$. Thus, if there were more than three neutrino families, H would increase and A would decrease with the same consequences as in the previous example. Similarly, if the number of neutrinos were very different from the number of antineutrinos, contrary to the assumptions in standard Big Bang cosmology, H would also increase.

Until 1990 the cosmic abundance of ^{4}He set the best limit to F_ν which was consistent with three to five. Now the very precise knowledge that F_ν is indeed 3 comes from the Large Electron–Positron (LEP) collider at CERN. One can summarize the Y_4 dependence on the parameters η, τ_n and F_ν in an empirical formula

$$Y_4 = 0.228 + 0.010 \ln\eta_{10} - 0.008\, \eta_{10}^{-2} + 0.012\,(F_\nu - 3) + 0.185 \left(\frac{\tau_n - 889.1}{889.1}\right), \quad (5.42)$$

where $\eta_{10} = 10^{10}\eta$, and τ_n should be expressed in seconds. In Fig. 31 we compare recent abundance determinations with the theoretical predictions for the η-dependence [11]. Taking the observational value of Y_4 to be 0.22–0.24 [11], the theoretical curve for Y_4 selects the range

$$1 \lesssim \eta_{10} \lesssim 3.3. \quad (5.43)$$

The combined D+^{3}He abundance is estimated to be less than 1.09×10^{-4} [11]. Since the theoretical curve in Fig. 31 is falling with increasing η, this gives a lower bound

$$\eta_{10} \geq 2.6. \quad (5.44)$$

The observational ^{7}Li/^{1}H ratio has been evaluated [12] to be in the range $(0.87 - 1.20) \times 10^{-10}$. Comparing this with the theoretical curve in Fig. 31, one obtains an allowed range of

$$2.0 \lesssim \eta_{10} \lesssim 3.6. \quad (5.45)$$

Thus all the observational evidence of nuclear abundances, F_ν, and τ_n are consistent with the Big Bang value in the range

$$2.8 \lesssim 10^{10}\, \eta \lesssim 4.2 \,. \quad (5.46)$$

It remains to convince oneself that the observed abundances are indeed of cosmological origin, and not significantly affected by later stellar processes. For instance, the deuterium abundance can easily be destroyed in later stellar events. The case of ^{7}Li is complicated because some fraction is due to later galactic cosmic ray spallation products. These have been estimated and subtracted to give the value in (5.45).

The helium isotopes ^{3}He and ^{4}He cannot be destroyed easily but they are continuously produced in stellar interiors. Some recent helium is blown off from supernova progenitors, but that fraction can be corrected for by observing the total abundance in hydrogen clouds of different age, and extrapolating it to time zero. The remainder is then primordial helium emanating from Big Bang nucleosynthesis.

From the nucleosynthesis value of η one can obtain an estimate of the baryon density parameter Ω_B. Combining Eqs. (5.24), (5.25) and $N_\gamma = 410$ photons/cm^3 from Eq. (5.11), this is

$$\Omega_B = \frac{N_B m_B}{\rho_c} = \frac{N_\gamma m_B}{\rho_c}\eta \simeq 3.65 \times 10^7 \eta h^{-2} \,, \quad (5.47)$$

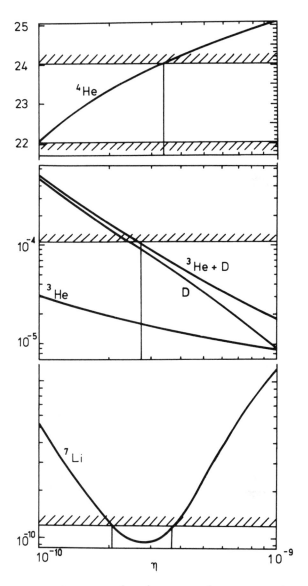

Fig. 31 Mass abundances of the nuclei ^{4}He, ^{3}He, D and ^{7}Li relative to ^{1}H, as functions of the baryon to photon ratio η. The curves are the predicted primordial nucleosynthesis abundances; the shaded regions are excluded by observations. Reproduced from references [10] and [11] by permission of K. Olive

where m_B is the proton mass. It then follows from the limits on η in (5.46) that

$$0.010 \lesssim \Omega_B h^2 \lesssim 0.015 .$$

Note that h ranges from 0.5 to 0.85 which implies values of Ω_B in the range

$$0.01 \lesssim \Omega_B \lesssim 0.05 . \tag{5.48}$$

Recall that direct observations of luminous (and therefore baryonic) matter quoted in Sec. 3.4 ranged between 5% and 45% , and that the energy density due to radiation was only $\Omega_{r,0} \simeq 4.17 \times 10^{-5}h^{-2}$. Thus we arrive at the very important conclusion that *there is too little baryonic matter to close the Universe*! Either the Universe is then indeed open, or there must exist other non-baryonic matter.

In Table 6 various derived properties of the present Universe are collected. In Fig. 32 the history of the Universe is summarized in nomograms relating the scales of temperature, energy, size, density and time. Note that so far we have only covered the events which occurred between 10^{13} K and 10^3 K.

Nuclear synthesis also goes on inside stars where the gravitational contraction increases the pressure and temperature so that the fusion process does not stop with helium. Our Sun is burning hydrogen to helium, which lasts about 10^{10} yr, a figure which is very dependent on the mass of the star. After that, helium burns to carbon in typically 10^6 years, carbon to oxygen and neon in 10^4 years, those to silicon in 10 years, and silicon to iron in 10 hours, whereafter the fusion chain stops. The birth of the rest of the elements had to wait for supernova explosions.

Table 6   Present properties of the Universe

Unit	Symbol	Value
Age	t_0	(14–18) Gyr
Mass	M_U	$\approx 10^{22}\ M_\odot$
Critical density	ρ_C	10.6 h^2 GeV/m^3
Mean luminous density	ρ_0	$\approx 10^{-27}$ kg/m^3
CMB radiation temperature	T_0	2.726 K
Cosmic neutrino temperature	T_ν	1.946 K
Radiation energy density	$\varepsilon_{r,0}$	1.40×10^{-12} erg/cm^3
Radiation density parameter	$\Omega_{r,0}$	$4.11\times 10^{-5}h^{-2}$
Entropy density	s/V	2888 cm^{-3}
CMB photon number density	N_γ	410 photons/cm^3
CMB temperature anisotropy	$\delta\,T/T$	1.1×10^{-5}
Cosmological constant	$|\lambda|$	$\lesssim 4\times 10^{-52}$ m^{-2}
Schwarzschild radius	$r_{C,\text{Universe}}$	$\gtrsim 11$ Gpc
^{4}He to ^{1}H mass ratio	Y_4	$0.22 - 0.24$
Baryon to photon ratio	η	$(2.8-4.2)\times 10^{-10}$
Baryon density parameter	Ω_B	0.01–0.05
Total density parameter	Ω_0	0.5−1.1
Deceleration parameter	q_0	≈ 0.5

113

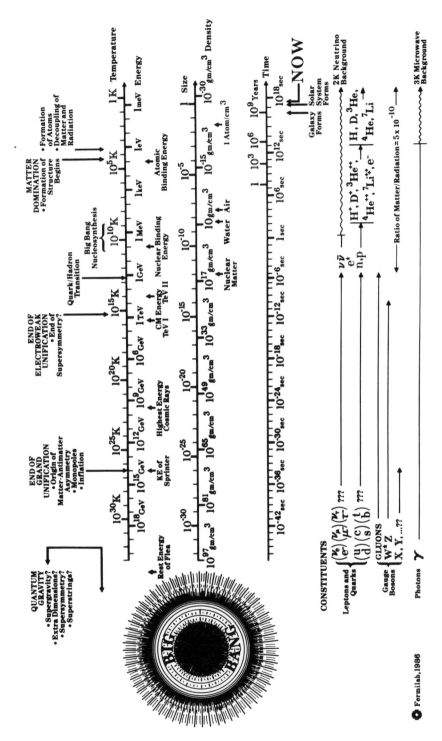

Fig. 32 The 'complete' history of the Universe, reproduced from reference [3] by permission of E. W. Kolb and M. S. Turner

Problems

1. Derive an equation for T_{eq} from the condition (5.3).
2. Use Wien's constant Eq. (4.77) and the CMB temperature to determine the wavelength of the CMB.
3. Use the present radiation energy density to calculate the pressure due to radiation in the Universe. Compare with the pressure due to a gas of galaxies calculated in Problem 3 of Chapter 4.
4. Show that an observer moving with velocity β in a direction θ relative to the CMB sees the rest frame blackbody spectrum with temperature T as a blackbody spectrum with temperature

$$T' = \frac{T}{\gamma(1 - \beta \cos \theta)} . \tag{5.49}$$

   To first order in β this gives the dipole anisotropy Eq. (5.12) [1].
5. The dipole anisotropy is measured to be $1.2 \times 10^{-3} T_0$. Derive the velocity of the Earth relative to the comoving coordinate system.
6. What is the size of density fluctuations required by Eq. (5.16) to produce the 'Great Wall' of length 170 h^{-1} Mpc, given the South Pole experiment anisotropy upper limit (5.20) ?
7. The rate of ν_e neutrino interactions Γ_ν for elastic scattering on electrons is proportional to the temperature squared. Calculate the proportionality factor, using T_0 and the knowledge that the Hubble expansion rate (4.60) was equal to the interaction at decoupling when the energy of the neutrinos was 2.3 MeV. Compare with the proportionality factor in Eq. (4.62).
8. Use the equation of relativistic kinematics (2.54) and conservation of four-momentum to calculate the energy of the photon liberated in Eq. (5.34), assuming that the ^{4}He nucleus is produced at rest.
9. The mean time for neutron capture by a proton is given by

$$t = (N_p \sigma v)^{-1} , \tag{5.50}$$

   where the capture cross section is $\sigma \simeq 10^{-28}$ cm^2. Compare this time with the neutron mean life at a temperature of $T = 10^9$ K when the baryon density is about 2×10^{-5} g/cm^3 [13].
10. Free nucleons are favoured over deuterons down to a radiation energy of 0.07 MeV. What is the ratio of photons with energies exceeding the deuteron binding energy 2.22 MeV to the number of protons at 0.07 MeV ?
11. Propose a two-stage fusion process leading to the production of ^{12}C.
12. Gamow's penetration factor (5.37) gives a rough idea about the ignition temperatures in stellar interiors for each fusion reaction. Estimate these under the simplifying assumption that the burning rates during the different periods are inversely proportional to the time spans (given at the end of this chapter), respectively. Take the hydrogen burning temperature to be 10^4 K.

References

1. P. J. E. Peebles, *Principles of Physical Cosmology*. Princeton University Press, Princeton, New Jersey, 1993.
2. J. C. Mather, E. S. Cheng, R. E. Eplee *et al., The Astrophysical Journal Letters*, **354** (1990) L37.
3. J. C. Mather, E. S. Cheng, D. A. Cottingham *et al., The Astrophysical Journal*, 1993, in press.
4. E. W. Kolb and M. S. Turner, *The Early Universe*, Addison-Wesley Publ. Co., Reading, Mass., 1990.
5. N. Kaiser and J. Silk, *Nature* **324** (1986) 529.
6. G. F. Smoot, C. L. Bennett, A. Kogut *et al., The Astrophysical Journal*, **396** (1992) L1.
7. T. Gaier, J. Schuster, J. Gundersen *et al., The Astrophysical Journal*, **398** (1992) L1 and preliminary report by J. Silk in *Nature* **356** (1992) 741.
8. S. S. Meyer, E. S. Cheng and L. A. Page, *The Astrophysical Journal*, *371* (1991) L7.
9. R. A. Watson, C. M. Gutierrez de la Cruz, R. D. Davies *et al., Nature*, **357** (1992) 660.
10. D. Lynden-Bell, S. M. Faber, D. Burstein *et al., The Astrophysical Journal*, **326** (1988) 19.
11. M. White, Physical Review, **D46**(1993) 4198; R. Crittenden, J. R. Bond, R. L. Davis, G. Efstathiou and P. J. Steinhardt, *Physical Review Letters*, **71** (1993) 324.
12. K. A. Olive, D. N. Schramm, G. Steigman and T. P. Walker, *Physics Letters B*, **236** (1990) 454.
13. K. A. Olive and D. N. Schramm, *Nature*, **360** (1992) 439.
14. F. H. Shu, *The Physical Universe*, University Science Books, Mill Valley, CA, 1982.

6 Particles and Symmetries

Prior to the time when electroweak reactions began to dominate, the Universe was dominated by different particles and interactions than those we have met so far. During the different epochs or phases the interactions of the particles were characterized by various symmetries governing their interactions. At each phase transition the symmetries and the physics changed radically. Thus we can only understand the electroweak epoch if we know where it came from and why. We must therefore start a journey backwards in time, where the uncertainties increase at each phase transition.

A very important symmetry is SU(2), which one usually meets for the first time in the context of electron spin. Even if electron spin is not an end to this chapter, it is a good and perhaps familiar introduction to SU(2). Thus we shall devote Section 6.1 to an elementary introduction to spin space, without the intention of actually carrying out spinor calculations.

Armed with SU(2) algebra we study three cases of SU(2) symmetry in particle physics: the isospin symmetry of the nucleons and the weak isospin symmetry of the leptons in Section 6.2, and weak isospin symmetry of the quarks in Section 6.3. We are then ready for the colour degree of freedom of quarks and gluons and the corresponding colour symmetry $SU(3)_c$. This leads up to the 'standard model' of particle physics which exhibits $SU(3)_c \otimes SU(2)_w \otimes U(1)_{B-L}$ symmetry.

In Section 6.4 we study the discrete symmetries of parity P, charge conjugation C and time reversal invariance T which will be of importance later in connection with the baryon–antibaryon asymmetry of the Universe.

In our present matter-dominated world all the above symmetries are more or less broken, with exception only for the colour symmetry $SU(3)_c$ and the combined discrete symmetry CPT. This does not mean that symmetries are unimportant, rather that the mechanisms of spontaneous symmetry breaking deserve attention. We take care of this in Section 6.5.

In Section 6.6 we assemble all our knowledge of particle symmetries and their spontaneous breaking in an attempt to describe the phase transitions in the primeval universe.

6.1 Spin Space

In Chapter 4 we introduced the quantal concept of spin. We found that the number n_{spin} of spin states of a particle was one factor contributing to its degrees of freedom g in thermal equilibrium, see Eq. (4.39). Thus horizontally and vertically polarized photons were counted as two effectively different particle species, although their physical properties were otherwise identical.

Charged particles with spin exhibit magnetic moment. In a magnetic field free electrons orient themselves as if they were little magnets. The classical picture that comes to mind is an electric charge spinning around an axis parallel to the external magnetic field. Thus the resulting magnetic moment would point parallel or antiparallel to the external field depending on whether the charge spinned right- or left-handedly with respect to the field. Although it must be emphasized that spin is an entirely quantal effect and not at all due to a spinning charge, the classical picture is helpful because the magnetic moment of the electron—whatever the mechanism for its generation—couples to the external field just like a classical magnet.

Let us take the spin axis of the electron to be given by the *spin vector* **S** with components S_x, S_y, S_z, and let the external magnetic field **B** be oriented in the z-direction,

$$\mathbf{B} = (0, 0, B_z) . \tag{6.1}$$

Then the potential energy due to the magnetic coupling of the electron to the external field is the *Hamiltonian operator*

$$H = A \ \mathbf{S} \cdot \mathbf{B} = A \ S_z B_z , \tag{6.2}$$

where A is a constant.

Measurements of H show that S_z is *quantized*, and that it can have only two values. With a suitable choice of units, these values are $\hbar/2$ and $-\hbar/2$. For convenience, one may replace **S** by a vector $\boldsymbol{\sigma}$ such that

$$\mathbf{S} = \tfrac{1}{2} \hbar \boldsymbol{\sigma} . \tag{6.3}$$

Then the two possible spin values of the electron correspond to

$$\sigma_z = \pm 1 . \tag{6.4}$$

This fits the classical picture insofar as the opposite signs would correspond to righthandedly and lefthandedly spinning charges, respectively. But the classical picture does not lead to quantized values, it permits a continuum of values. One consequence of the quantum dichotomy is that free electrons in thermal equilibrium with radiation contribute each $n_{spin} = 2$ degrees of freedom to g.

The above conclusions followed when the external magnetic field **B** was turned on. What if **B**=0, where do the electron spin vectors $\boldsymbol{\sigma}$ point then? The answer of course is 'anywhere', but this we cannot confirm experimentally, because we precisely need a non-vanishing external field to measure H or σ_z. Thus quantum mechanics leads to a paradoxical situation: even if electrons which are not subject to observation or magnetic influence may have their spins arbitrarily oriented; by observing them we force them into either one of the two states (6.4).

Quantum mechanics resolves this paradox by introducing statistical laws which govern averages of ensembles of particles, but which say nothing about the individual events. Consider the free electron *before* the field B_z has been switched on. Then it can be described as having the probability P to yield the σ_z value 1 *after* the field has been switched on, and the probability $1 - P$ to yield the value –1. The value of P is anywhere between 0 and 1, as the definition of probability requires.

After a measurement has yielded the value $\sigma_z = 1$, a subsequent measurement has the probability 1 to yield 1 and 0 to yield –1. Thus a measurement changes the *spin state* from indefinite to definite. The definite spin states are characterized by σ_z being ± 1; the indefinite state is a *superposition* of these two states. This can be formulated either geometrically or algebraically. Let us turn to the geometrical formulation first.

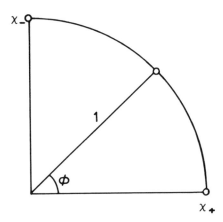

Fig. 33   Two-component spin space

Consider an abstract two-dimensional space with orthogonal axes labelled χ_+ and χ_-, see Fig. 33. Let us draw an arc through the points $\chi_+ = 1$ and $\chi_- = 1$. The length of the radius vector χ is then 1 for arbitrary polar angles ϕ, and its coordinates are $(\cos \phi, \sin \phi)$. Let us identify the square of the χ_+ coordinate, $\cos^2 \phi$, with the probability P. It then follows that $1 - P = \sin^2 \phi$.

We now identify every possible spin state with a vector from the origin to a point on the arc in Fig. 33. These vectors in spin space are called *spinors* to distinguish them from the spin vector σ in ordinary space. Let us write the χ_+ and χ_- coordinates in column form. Then the points

$$\chi_+ = \begin{pmatrix} 1 \\ 0 \end{pmatrix}, \chi_- = \begin{pmatrix} 0 \\ 1 \end{pmatrix} \tag{6.5}$$

are spinors corresponding to $\sigma_z = 1$ and $\sigma_z = -1$, respectively. An arbitrary point on the arc with coordinates $\cos\phi$, $\sin\phi$ corresponds to a linear superposition of the $\sigma_z = 1$ and $\sigma_z = -1$ states. Using the spinors (6.5) as base vectors, this can clearly be written

$$\chi = \cos\phi \begin{pmatrix} 1 \\ 0 \end{pmatrix} + \sin\phi \begin{pmatrix} 0 \\ 1 \end{pmatrix} . \tag{6.6}$$

To summarize, points on the arc correspond to states of the electron before any spin measurement has been done, and the points (6.5) correspond to *prepared states* after a measurement. Points elsewhere in the (χ_+, χ_-)-plane have no physical interpretation. The coordinates of points on the arc have no direct physical meaning, but their squares correspond to the probabilities of the outcome of a subsequent spin measurement.

The spin space has as many dimensions as there are possible outcomes, in the electron case two. It is an abstract space spanned by spinors χ, not by the spin vectors **S** or $\boldsymbol{\sigma}$ of ordinary three-dimensional space. The spinor formalism allows the two spin states of the electron to be treated symmetrically: the electron is just a single two-component object, and both components have identical properties in all other respects.

Let us next turn to the algebra of spin states. The rotation of a spinor χ_1 into another spinor χ_2 corresponds algebraically to an operator U operating on χ_1,

$$\chi_2 = U\chi_1 . \tag{6.7}$$

Both components of χ_1 are then transformed by U, thus U in component notation must be described by a 2×2 matrix. This is most easily exemplified by the matrix S_+ which rotates χ_- into χ_+,

$$S_+\chi_- = \begin{pmatrix} 0 & 1 \\ 0 & 0 \end{pmatrix} \begin{pmatrix} 0 \\ 1 \end{pmatrix} = \begin{pmatrix} 1 \\ 0 \end{pmatrix} = \chi_+, \tag{6.8}$$

and the matrix S_- which does the opposite. The operators S_+ and S_- are called *raising* and *lowering operators*, respectively.

We noted above that the weights $\cos\phi$ and $\sin\phi$ in Eq. (6.6) have no direct physical interpretation, but that their squares are real numbers with values between 0 and 1. We may as well abandon the geometrical idea that the weights are real numbers, since no physical argument requires this. Let us therefore replace them by the complex numbers $a_+ = ae^{i\alpha}$ and $a_- = be^{i\beta}$. The magnitudes a,b are real numbers which can be required to obey the same relations as $\cos\phi$ and $\sin\phi$,

$$|a_+|^2 + |a_-|^2 = a^2 + b^2 = 1 , \quad 0 \le |a_+| \le 1, \quad 0 \le |a_-| \le 1 , \tag{6.9}$$

but the phase angles α and β need not have any physical interpretation. Thus we can make the identifications

$$P = |a_+|^2 , \quad 1 - P = |a_-|^2 . \tag{6.10}$$

The change from real weights in Eq. (6.6) to complex weights does not at first sight change the physics, but it adds some useful freedom to the theory. Quantum mechanics embodies this in the important *principle of superposition* which states that *if the spinors* χ_+ *and* χ_- *describe physical states, then every linear superposition of them with complex coefficients* $a_\pm$,

$$\chi = a_+\chi_+ + a_-\chi_- \, , \tag{6.11}$$

also describes a physical state.

It follows from the complexity of the $a_\pm$ that also the matrix U in Eq. (6.7) is complex. Moreover, U is restricted to transform a point on the unit circle in Fig. 33 into another point on the circle. One then proves easily that the operator U must be *unitary*, obeying

$$UU^\dagger = U^\dagger U = 1 \, , \tag{6.12}$$

where the superscript † implies transposition and complex conjugation. Actually the unitarity condition (6.12) follows from Eqs. (6.7) and (6.9).

If the spin space were one-dimensional the unitarity condition (6.12) would be satisfied by any pure *phase transformation*

$$U = e^{i\theta \cdot 1} \, . \tag{6.13}$$

The number 1 is of course superfluous in the exponent, but we keep it for later reference. All operators of this form are elements of a mathematical *group* called U(1). Here U stands for unitary and (1) for the *order* of the group. The phase angle θ is a *real parameter of the group* having a global value all over spacetime, that is, it does not depend on the spacetime coordinates. The transformation (6.13) is therefore called a *global gauge transformation*.

A similar situation occurs in another familiar context: Maxwell's equations for the electromagnetic field are invariant under a phase transformation

$$U = e^{iQ\theta(x)} \, , \tag{6.14}$$

where the electric charge Q is a conserved quantity. Now, however, the parameter of the group is the product $Q\theta(x)$ which depends on the local spacetime coordinate x through the function $\theta(x)$. The U(1) symmetry implies that Maxwell's equations are independent of the local choice of $\theta(x)$. The transformation (6.14) is then called a *local gauge transformation* (see e.g. [1]). Because this gauge symmetry is exact, the *gauge boson* of the theory, the photon, is exactly massless.

This situation is quite similar to the principle of covariance which demanded that the laws of physics should be independent of the choice of local spacetime coordinates. The principle of covariance is now replaced by the *gauge principle* which applies to gauge field theories.

In the spin space of the electron the operators U are represented by unitary 2×2 matrices which in addition obey the special condition

$$\det U = 1. \tag{6.15}$$

This defines them to be elements of the global group SU(2), of order two. The letter S in SU(2) stands for Special condition. In group theory parlance the two-component spinors in Eq. (6.5) are *doublet representations of SU(2)*.

It is possible to express the U operators in terms of exponentiations,

$$U = e^{iH} , \qquad (6.16)$$

analogously to the one-dimensional case (6.13). This requires the quantities H to be complex 2×2 matrices as well. Substituting the expression (6.16) into the unitarity condition (6.12) we have

$$UU^{\dagger} = e^{i(H-H^{\dagger})} = 1 . \qquad (6.17)$$

It follows that the operators H must be *Hermitian*,

$$H=H^{\dagger} . \qquad (6.18)$$

Moreover, the special condition (6.15) requires the H matrices to be traceless. The most general way to write a 2×2 Hermitian matrix is

$$H = \theta_1 \sigma_x + \theta_2 \sigma_y + \theta_3 \sigma_z , \qquad (6.19)$$

where the σ_i are the *Pauli matrices*

$$\sigma_x = \begin{pmatrix} 0 & 1 \\ 1 & 0 \end{pmatrix} , \ \sigma_y = \begin{pmatrix} 0 & -i \\ i & 0 \end{pmatrix} , \ \sigma_z = \begin{pmatrix} 1 & 0 \\ 0 & -1 \end{pmatrix} . \qquad (6.20)$$

Note that all the σ_i are traceless, and only σ_z is diagonal.

Comparing the exponent in Eq. (6.13) with the expression (6.19) we see that the single parameter θ in the one-dimensional case corresponds to three real parameters θ_i in SU(2). The superfluous number 1 in U(1) is the vestige of the Pauli matrices appearing in SU(2). It shares with them the property of having the square 1.

The number 1 generates the ordinary algebra—a trivial statement, indeed !—whereas the Pauli matrices generate a new, *noncommutative algebra*. In *commutative* or *Abelian algebras* the product of two elements θ_1 and θ_2 can be formed in either order,

$$\theta_1 \theta_2 - \theta_2 \theta_1 \equiv [\theta_1, \theta_2] = 0 .$$

Here the square bracket expression is called a *commutator*. In the *non-Abelian* algebra SU(2) the commutator of two elements does not in general vanish. For instance, the commutator of two Pauli matrices σ_i is

$$[\sigma_i, \sigma_j] = 2i\sigma_k , \qquad (6.21)$$

where i,j,k represent any cyclic permutation of x,y,z. The *anticommutator*, defined by

$$\{\sigma_i, \sigma_j\} \equiv \sigma_i \sigma_j + \sigma_j \sigma_i \qquad (6.22)$$

can be shown to vanish for any pair of Pauli matrices.

The quantum theory is purporting to describe the *observation* of properties of *physical states*, and the possible numerical *outcome* of such observations. For instance, the observation of the spin of an electron is described by an *operator* S acting on a *spin state* χ_1, and the numerical *outcome* is then *s*. This can be formulated as in Eq. (6.7),

$$S\chi_1 = s\chi_2 \ . \tag{6.23}$$

An operator of the form (6.19) mixes in general the components of χ_1, and moves χ_1 onto another point χ_2 on the arc.

An exception occurs when the operator in Eq. (6.23) is a diagonal matrix like $S_z = \frac{1}{2}\hbar\sigma_z$, because it does not mix the upper and lower components of the spinor. Physically of course the operation with S_z implies a measurement of the spin in the *z* direction (when a field B_z is applied). If the electron at the time of the measurement is in either one of the *pure states* χ_+ or χ_-, Eq. (6.23) becomes, respectively,

$$S_z\chi_+ = \tfrac{1}{2}\hbar \begin{pmatrix} 1 & 0 \\ 0 & -1 \end{pmatrix} \begin{pmatrix} 1 \\ 0 \end{pmatrix} = \tfrac{1}{2}\hbar \begin{pmatrix} 1 \\ 0 \end{pmatrix} = \tfrac{1}{2}\hbar\chi_+ \ , \tag{6.24}$$

$$S_z\chi_+ = \tfrac{1}{2}\hbar \begin{pmatrix} 1 & 0 \\ 0 & -1 \end{pmatrix} \begin{pmatrix} 0 \\ 1 \end{pmatrix} = -\tfrac{1}{2}\hbar \begin{pmatrix} 0 \\ 1 \end{pmatrix} = -\tfrac{1}{2}\hbar\chi_- \ . \tag{6.25}$$

Thus the outcome of the observation appears as the numerical factor $s_z = \pm\frac{1}{2}\hbar$ in front of the unchanged spinor $\chi_\pm$ on the right, respectively. An equation of this kind is called an *eigenvalue equation*, the spinors $\chi_\pm$ are called *eigenstates of the operator* S_z, and the numerical values s_z are its *eigenvalues*.

Note that Eq. (6.23) is not an eigenvalue equation when $\chi_1 \neq \chi_2$. For instance, the linear superposition (6.11) of pure states does not satisfy an eigenvalue equation with eigenvalue *s*,

$$S_z\chi = S_z(a_+\chi_+ + a_-\chi_-) = \tfrac{1}{2}\hbar(a_+\chi_+ - a_-\chi_-) \neq s\chi \ .$$

The important lesson of this is that *possible observations are operators* which can be represented by *diagonal matrices*, and the numbers appearing on the *diagonal* are the *possible outcomes* of the observation. Moreover, the operators must be linear since they operate in a linear space, transforming spinors into spinors.

In ordinary space the spin vector **S** has a length which of course is a real positive number. Since its projection on the z axis is either $s_z = +\frac{1}{2}\hbar$ or $s_z = -\frac{1}{2}\hbar$, the length of **S** must be $S \equiv |\mathbf{S}| = \frac{1}{2}\hbar$. The sign of s_z indicates that **S** is parallel or antiparallel to the z direction, respectively. Consider a system formed by two electrons *a* and *b* with spin vectors $\mathbf{S}_a$ and $\mathbf{S}_b$ and spinor states $\chi_+^a, \chi_-^a, \chi_+^b, \chi_-^b$. The sum vector

$$\mathbf{S} = \mathbf{S}_a + \mathbf{S}_b$$

can clearly take any values in the continuum between 0 and $1\hbar$ depending on the relative orientations. However, quantum mechanics requires *S* to be quantized to integral multiples of $\hbar$, in this case to 0 or 1. This has important consequences for atomic spectroscopy and particle physics.

Let us study the projections s_z of **S** using the notation $|S,s_z\rangle$ for *two-electron spin states*, and dropping $\hbar$. When $S = 0$ the projection can of course only be $s_z = 0$. Then the two electrons form a *spinless state* $|0,0\rangle$. When $S = 1$ the projection can be $+1$ or -1, but also 0 because quantum mechanics requires that also s_z takes on all possible integral values. The three possible states are then

$$|1,1\rangle , \quad |1,0\rangle , \quad |1,-1\rangle . \tag{6.26}$$

It is easy to identify two of these two-electron states with the corresponding combination of one-electron states,

$$|1,1\rangle = \chi_+^a \chi_+^b , \quad |1,-1\rangle = \chi_-^a \chi_-^b . \tag{6.27}$$

However, the two states $|1,0\rangle$ and $|0,0\rangle$ having $s_z = 0$, could both correspond to either one of the two combinations

$$\chi_+^a \chi_-^b , \quad \chi_-^a \chi_+^b , \tag{6.28}$$

but it is not obvious which corresponds to which.

To resolve this problem we must make use of a symmetry called *exchange symmetry*. Physics must be invariant under the exchange of two *identical particles*. In the expressions (6.27) the two spinors χ enter symmetrically, thus by exchanging the labels a and b of the identical electrons we get the same products back. Hence these two products are *symmetric under exchange*. It is reasonable to require that all the three $S = 1$ states (6.26) should have the same symmetry.

The two independent products (6.28) are neither symmetric nor antisymmetric. Let us form two linear combinations of them, one symmetric and one antisymmetric. We then identify $|1,0\rangle$ with the symmetric combination

$$|1,0\rangle = \sqrt{\tfrac{1}{2}}(\chi_+^a \chi_-^b + \chi_-^a \chi_+^b) , \tag{6.29}$$

and the spinless state with the antisymmetric combination

$$|0,0\rangle = \sqrt{\tfrac{1}{2}}(\chi_+^a \chi_-^b - \chi_-^a \chi_+^b) . \tag{6.30}$$

The physical meaning of the linear superpositions (6.29) and (6.30) is the following. When the two-electron system is in either one of these states, an observation of the spins of the individual electrons a and b would result with equal probability P in $\chi_+^a \chi_-^b$ (electron a has spin $+\tfrac{1}{2}\hbar$ and electron b has spin $-\tfrac{1}{2}\hbar$) and in $\chi_-^a \chi_+^b$ (the reversed case). There are no other possibilities, and the sum of the probabilities for all possibilities must be 1, hence the values $P = \tfrac{1}{2}$. The absolute squares of the coefficients in the expressions (6.29) and (6.30) are interpreted as P, hence their absolute values must be $\sqrt{\tfrac{1}{2}}$. With this choice all the four states (6.27), (6.29) and (6.30) are represented by vectors of equal length in the spin space of $\chi^a \chi^b$. The signs of the coefficients are unphysical and unobservable in the experiment described, but one can design other types of experiments where the sign becomes an observable.

6.2 SU(2) Symmetries

The proton and the neutron are very similar, except in their electromagnetic properties: they have different charges and magnetic moments. Suppose one could switch off the electromagnetic interaction, leaving only the *strong interaction* at play. Then the p and n fields would be identical, except for their very small mass difference. Even that mass difference one could explain as an electromagnetic effect, because it is of the expected order of magnitude.

Making use of SU(2) algebra one would then treat the nucleon N as a two-component state in an abstract *charge space*,

$$N = \begin{pmatrix} p \\ n \end{pmatrix} . \tag{6.31}$$

The p and n fields are the base vectors spanning this space,

$$p = \begin{pmatrix} 1 \\ 0 \end{pmatrix}, \quad n = \begin{pmatrix} 0 \\ 1 \end{pmatrix} . \tag{6.32}$$

In analogy with the spin case, these states are the eigenstates of an operator, here denoted $I_3 = \frac{1}{2}\sigma_z$, with eigenvalues

$$I_3 = \pm\tfrac{1}{2} . \tag{6.33}$$

Thus the proton with charge $Q = +1$ has $I_3 = +\frac{1}{2}$, and the neutron with $Q = 0$ has $I_3 = -\frac{1}{2}$. It is then convenient to give I_3 physical meaning by relating it to charge,

$$Q = \tfrac{1}{2} + I_3 . \tag{6.34}$$

It follows that the *charge operator* in matrix form is

$$Q = \tfrac{1}{2} \begin{pmatrix} 1 & 0 \\ 0 & 1 \end{pmatrix} + \tfrac{1}{2} \begin{pmatrix} 1 & 0 \\ 0 & -1 \end{pmatrix} = \begin{pmatrix} 1 & 0 \\ 0 & 0 \end{pmatrix} , \tag{6.35}$$

where the charges of p and n, respectively, appear on the diagonal.

One can also define two operators $I_1 = \frac{1}{2}\sigma_x$ and $I_2 = \frac{1}{2}\sigma_y$ in order to recover the complete SU(2) algebra. These operators interchange the charge states, for instance

$$I_1 N = \tfrac{1}{2} \begin{pmatrix} 0 & 1 \\ 1 & 0 \end{pmatrix} \begin{pmatrix} p \\ n \end{pmatrix} = \tfrac{1}{2} \begin{pmatrix} n \\ p \end{pmatrix} . \tag{6.36}$$

The I_1, I_2, I_3 are the components of the *isospin* vector $\mathbf{I}$ in an abstract three-dimensional space. This contrasts with the spin case where σ is a vector in ordinary three-dimensional space.

The advantage of this notation is that the strong interactions of protons and neutrons as well as any linear superposition of them are treated symmetrically. One says that strong interactions possess *isospin symmetry*. Just as in the case of electron spin, this is a global symmetry.

In Nature the isospin symmetry is not exact because one cannot switch off the electromagnetic interactions, as we supposed to begin with. Since the main asymmetry between the proton and the neutron is precisely expressed by their different electric charges, electromagnetic interactions are not isospin-symmetric. However, the strong interactions are so much stronger, as is witnessed by the fact that atomic nuclei containing large numbers of protons do not blow apart in spite of the Coulomb repulsion. Thus isospin symmetry is approximate, and it turns out to be a more useful tool in particle physics than in nuclear physics.

The lessons learned from spin space and isospin symmetry have turned out to be useful in still other contexts. As we have seen in Section 4.3, the electron and its neutrino form a family characterized by the conserved *e*-lepton number L_e. They participate in similar electroweak reactions, but they differ in mass and in their electromagnetic properties, as do the proton and neutron. The neutrino may even be massless, so the mass difference is too important to be blamed on their different electric charges.

In spite of these asymmetries, it turns out to be a fruitful approach to consider the leptons to be components of three SU(2) doublets,

$$\ell_e = \begin{pmatrix} \nu_e \\ e^- \end{pmatrix} , \quad \ell_\mu = \begin{pmatrix} \nu_\mu \\ \mu^- \end{pmatrix} , \quad \ell_\tau = \begin{pmatrix} \nu_\tau \\ \tau^- \end{pmatrix} . \tag{6.37}$$

Their antiparticles form three similar doublets where the up/down order is reversed. Thus we meet yet another abstract space, *weak isospin space*, which is not identical to either spin space or isospin space; only the algebraic structure is the same. This space is spanned by spinorlike base vectors

$$\nu_e = \begin{pmatrix} 1 \\ 0 \end{pmatrix} , \quad e^- = \begin{pmatrix} 0 \\ 1 \end{pmatrix} . \tag{6.38}$$

Making use of the notation (6.37) the reactions (4.30), (4.35), and (4.36) can be written as one,

$$\overset{(-)}{\ell}_i + \overset{(-)}{\ell}_j \to \overset{(-)}{\ell}_i + \overset{(-)}{\ell}_j , \tag{6.39}$$

where the subscripts *i* and *j* refer to *e*, μ or τ. The reactions (4.37) and (4.38) can then be summarized by

$$\ell_i + \overset{(-)}{\ell}_i \to \ell_j + \overset{(-)}{\ell}_j . \tag{6.40}$$

In analogy with the spin and isospin cases, the states (6.38) are the eigenstates of an operator

$$T_3 = \tfrac{1}{2}\sigma_z = \tfrac{1}{2} \begin{pmatrix} 1 & 0 \\ 0 & -1 \end{pmatrix} . \tag{6.41}$$

The eigenvalues $T_3 = \pm\tfrac{1}{2}$ appearing on the diagonal are called *weak charge*. One can also define a weak isospin vector **T** with components T_1, T_2, T_3 spanning an abstract three-dimensional space.

In the isospin case we had to 'switch off' electromagnetism in order to achieve a global symmetry between the differently charged p and n states of the nucleon. In the case of the electroweak theory, the particle states are represented by gauge fields which are locally gauged (see, for example, ref. [1]). In addition a trick has been invented which incorporates both charge and weak charge in the symmetry group. This trick is to enlarge the gauge symmetry to the direct product of two local gauge symmetry groups,

$$SU(2)_w \otimes U(1)_Y \ .$$

Here w stands for weak isospin, and Y is a new quantum number called *weak hypercharge*, the parameter of a U(1) group. If one defines the latter by

$$\tfrac{1}{2}Y = Q - T_3 \ , \tag{6.42}$$

all the leptons have $Y = -1$ regardless of charge, and all the antileptons have $Y = 1$.

The assumption that Nature observes $SU(2)_w \otimes U(1)_Y$ symmetry implies that the electroweak interaction does not see any difference between the neutrino and the electron fields and linear superpositions of them,

$$\ell_e = a_+\nu_e + a_-e^- \ ;$$

they all have the same weak hypercharge and the same L_e. That the symmetry is a local gauge symmetry implies that the laws of electroweak interactions are independent of the 'local' choice of gauge functions, analogous to $\theta(x)$ in Eq. (6.14). However, Nature does not realize this symmetry exactly; it is a *broken symmetry*. This is seen by the fact that of the four gauge bosons γ, Z^0, W^+, W^- mediating the electroweak interaction, three are massive. In an exact local-gauge symmetric theory all gauge bosons are massless, as is the photon in the case of the exact $U(1)$.

Actually, the electroweak force is the outcome of a long and difficult search to unify the forces of Nature. The first milestone on this road was set up by Maxwell when he managed to unify electricity and magnetism into one electromagnetic force. Since then Science has been concerned with four forces of Nature: the strong force responsible for the stability of nuclei, the electromagnetic force responsible for atomic structure and chemistry, the weak force which played such an important cosmological rôle during the late radiation era and the gravitational force acting during the matter-dominated era. All of these forces are described by local-gauge field theories.

Einstein attempted in vain to unify gravitation and electromagnetism during his last 20 years. A breakthrough came in 1967 when Sheldon Glashow, Steven Weinberg and Abdus Salam succeeded in unifying the weak and electromagnetic forces into the electroweak interaction. Since then the goal has been to achieve a *grand unified theory* (GUT) which would unify strong and electroweak interactions, and ultimately gravitation as well in a *theory of everything* (TOE).

6.3 Hadrons and Quarks

The spectrum of *hadrons* participating in strong interactions is extremely rich. The hadrons comprise two large classes of particles already encountered, the baryons and the mesons. Charged hadrons also have electromagnetic interactions, and all hadrons have weak interactions, but strong interactions dominate whenever possible.

The reaction rate of strong interactions exceeds by far the rate of other interactions. Weak decays like reactions (4.56–4.58) typically require mean lives ranging from microseconds to picoseconds. An electromagnetic decay like

$$\pi^0 \rightarrow \gamma + \gamma$$

takes place in less than 10^{-16} s, and heavier particles may decay 1000 times faster. Strongly decaying hadrons, however, have mean lives of the order of 10^{-23} s.

Some simplification of the hadron spectrum may be achieved by introducing isospin symmetry. Then the nucleon N stands for n and p, the pion π stands for π^+, π^0, π^-, the *kaon K* for the *strange mesons* K^+, K^0 with mass 495 MeV, etc. However, it was realized in 1962 that the hadron spectrum possessed more symmetry than that. In fact, Murray Gell-Mann and George Zweig showed that all hadrons known could be built out of three hypothetical states called *quarks*, $q = u, d, s$, which spanned a three-dimensional space with SU(3) gauge symmetry. This SU(3) group is an extension of the isospin SU(2) group to include *strangeness*, an additive quantum number possessed by the kaon and many other hadrons. The *up* and *down* quark fields u, d form an isospin doublet like Eq. (6.31), whereas the *strange quark s* is an isospin-neutral singlet. Together they form the basic building block of SU(3), a triplet of quarks of three *flavours*.

In the quark model the mesons are $q\bar{q}$ bound states, and the baryons are qqq bound states. The differences in hadron properties can be accounted for in two ways: the quarks can be excited to higher angular momenta, and in addition the three flavours offer various combinations. For instance, the nucleon states are the ground state configurations

$$p = uud , \quad n = udd. \tag{6.43}$$

The mesonic ground states are the pions and kaons with the configurations

$$\pi^+ = u\bar{d} , \quad \pi^0 = \frac{1}{\sqrt{2}}(u\bar{u} + d\bar{d}) , \quad \pi^- = d\bar{u} , \tag{6.44}$$

$$K^+ = u\bar{s} , \quad K^0 = d\bar{s} .$$

The quarks are fermions just like the leptons, and in spin space they are SU(2) doublets.

After the discovery in the sixties of the electroweak gauge symmetry $SU(2)_w \otimes U(1)_Y$ for the then known e- and μ-lepton families and the u, d family of quarks, it was realized that this could be the fundamental theory of electroweak interactions. But the theory clearly needed a fourth flavour quark to complete a second $SU(2)_w$ doublet together with the s-quark. To keep the s-quark as a singlet would not do: it would be T_3-neutral and not feel the weak interactions. In 1974 the long predicted *charmed quark c* was discovered simultaneously by two teams, an MIT team led by Sam Ting, and a SLAC team lead by Burt Richter.

The following year the issue was confused once more when another SLAC team led by Martin Perl discovered the τ lepton, showing that the lepton families were three. This triggered a search for the τ neutrino and the corresponding two quarks, if they existed.

In 1977 a team at the Cornell $e + e-$ collider CESAR led by Leon Lederman found the fifth quark with the same charge as the d and s, but with its own new flavour. It was therefore a candidate for the bottom position in the third quark doublet. Some physicists lacking the imagination of those who invented 'strangeness' and 'charm', baptised it *bottom quark b*, although the name *beauty* has also been used. The missing companion to the bottom quark in the third $SU(2)_w$ doublet was prosaically called *top quark, t*. The fields of the three quark families can then be ordered as

$$\begin{pmatrix} u \\ d \end{pmatrix}, \quad \begin{pmatrix} c \\ s \end{pmatrix}, \quad \begin{pmatrix} t \\ b \end{pmatrix}. \tag{6.45}$$

The top quark has not been discovered as of the time of writing, but several searches are going on.

The strong interaction symmetry which had started successfully with SU(3) for three quarks, would logically be enlarged to SU(n) for quarks of n flavours. However, the quark masses, although not directly measurable, are so vastly different that even SU(4) is a badly broken symmetry and not at all useful. Only the isospin SU(2) subgroup and the flavour SU(3) subgroup continue to be useful, in particular for the classification of hadrons.

It follows from the quark structure (6.43) of the nucleons that each quark possesses baryon number $B = \frac{1}{3}$. The $SU(2)_w$ symmetry requires all the $T_3 = \frac{1}{2}$ (upper) states in the doublets to have the same charge Q_u, and all the $T_3 = -\frac{1}{2}$ (lower) states to have the charge Q_d. To match the nucleon charges, the charges of the quarks have to satisfy the relations

$$2Q_u + Q_d = 1, \quad Q_u + 2Q_d = 0. \tag{6.46}$$

The solution chosen by Nature is

$$Q_u = \tfrac{2}{3}, \quad Q_d = -\tfrac{1}{3}. \tag{6.47}$$

It now follows from the definition of weak hypercharge Y in Eq. (6.42) that all the quarks have the same weak hypercharge, for instance

$$Y_d = -\tfrac{2}{3} + 1 = \tfrac{1}{3}. \tag{6.48}$$

Thus the quarks differ from the leptons in weak hypercharge. Actually we can get rid of the somewhat artificial notion of weak hypercharge by noting that

$$Y = B - L, \tag{6.49}$$

where B is the baryon number not possessed by leptons, and L is the lepton number not possessed by baryons. Thus for leptons $B - L = -1$ whereas for quarks it is $\frac{1}{3}$.

The electroweak interactions of the leptons and quarks are mediated by the *gauge bosons* γ, Z^0, W^+, W^-. Two examples of this interaction were illustrated by the Feynman diagrams in Figs. 20 and 21, where each line corresponds to a particle. Figure 34 shows the Feynman diagram for neutron β decay, reaction (4.65), where the decomposition into

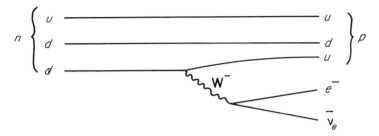

Fig. 34   Feynman diagram for neutron β decay with quark lines

quarks is explicit, a nucleon corresponding to three quark lines. As is seen, the decay of a neutron involves the transformation of a d quark of charge $-\frac{1}{3}$ into a u quark of charge $\frac{2}{3}$ and a virtual W^- boson. The final quark system is therefore that of a proton. Two of the quarks do not participate in the reaction at all, they remain *spectators* of what is going on. Subsequently the virtual W^- produces a lepton–antilepton pair, conserving electric charge and keeping the total lepton number at $L = 0$.

The strong interactions of hadrons which were never very well understood obviously had to be replaced by interactions at the quark level. For this a new mediator is needed, the *gluon*, which is a vector boson like the previously met mediators of interactions. The gluon is also then responsible for binding quarks into hadrons, but it does not feel the leptons.

The quarks have another intrinsic property which they do not share with the leptons. It appears from problems in hadron spectroscopy, and from the rates of certain hadronic reactions which occur three times faster than expected, that each quark actually must come in three versions. These versions do not differ from each other in any so far encountered respect, so they require a new property called *colour*. To distinguish quarks of different colour one may choose to call them red, blue and yellow (R,B,Y), for instance. They span an abstract three-dimensional space with $SU(3)_c$ symmetry (c for colour), and base vectors

$$q_R = \begin{pmatrix} 1 \\ 0 \\ 0 \end{pmatrix}, \quad q_B = \begin{pmatrix} 0 \\ 1 \\ 0 \end{pmatrix}, \quad q_Y = \begin{pmatrix} 0 \\ 0 \\ 1 \end{pmatrix}, \tag{6.50}$$

where q stands for the flavours u,d,c,s,t,b.

Colour is an absolutely conserved quantum number, in contrast to flavour which is conserved in strong and electromagnetic interactions, but broken in weak interactions. Since the gluon is interacting with quarks mediating the *colour force*, it must itself carry colour so that it can change the colour of a quark. The same situation occurs in electroweak interactions where the conservation of charge, as for instance in Fig. 34, requires the W to carry charge so that it can change a d quark into a u quark. Gluons interact with gluons because they possess colour charge, in contrast to photons which do not interact with photons because of their lack of electric charge.

Since there are quarks of three colours, there must exist nine gluons, one for each distinct colour pair. Thus the gluon colours are $B\bar{B}$, $B\bar{R}$, $B\bar{Y}$, $R\bar{B}$, $R\bar{R}$, $R\bar{Y}$, $Y\bar{B}$, $Y\bar{R}$, $Y\bar{Y}$. Actually group theory allows ony eight gluons of different hues to couple to quarks. The one excluded hue is a linear combination of $B\bar{B}$, $R\bar{R}$, and $Y\bar{Y}$ which is colour-neutral and totally antisymmetric under the exchange of any two colours, just like the two-electron state (6.30) was totally spin-neutral. Such states are *singlets* under the corresponding group, and the interaction is completely blind to them.

It is a curious fact that free quarks never appear in the laboratory, in spite of ingenious searches. When quarks were first invented to explain the spectroscopy of hadrons, they were thought to be mere abstractions without real existence. Their reality was doubted by many people since nobody had succeeded in observing them. But it was gradually understood that their non-observability was quite natural for deep reasons related to the properties of vacuum.

The quark and the antiquark in a meson are like the north and south poles of a magnet which cannot be separated, because there exists nothing such as a free magnetic north pole. If one breaks the bound of the poles, each of the pieces will become a new magnet with a north and a south pole. The new opposite poles are generated at the break, out of the vacuum so to speak . Similarly, if one tries to break a $q\bar{q}$ pair, a new $q\bar{q}$ pair will be generated out of the vacuum at the break, and one thus obtains two new $q\bar{q}$ mesons which are free to fly away.

To account for this property one must assign curious features to the potential responsible for the binding of the $q\bar{q}$ system: the larger the interquark distance, the stronger the potential. Inversely, at very small interquark distances the potential can be so weak that the quarks are essentially free ! This feature is called *asymptotic freedom*.

Since the colour property is not observed in hadrons, they must be colour-neutral. How can one construct colour-neutral compounds of coloured constituents? For the mesons which are quark–antiquark systems the answer is quite simple: for a given quark colour the antiquark must have the corresponding anti-colour. Thus for instance the π^+ meson is a $u\bar{d}$ system when we account for flavours only, but if we account also for colour three π^+ mesons are possible, corresponding to $u_B\bar{d}_B$, $u_R\bar{d}_R$ and $u_Y\bar{d}_Y$, respectively. Each of these is colour-neutral, so hadronic physics does not distinguish between them.

Also the baryons which are qqq states must be colour-neutral. This is possible for a totally antisymmetric linear combination of three quarks q,q',q'', having three colours each. Generalizing the expression (6.30) for the two-electron state, this linear combination is

$$\frac{1}{\sqrt{6}}(q_R q'_Y q''_B - q_R q'_B q''_Y + q_B q'_R q''_T - q_B q'_Y q''_R + q_Y q'_B q''_R - q_Y q'_R q''_B) \ . \qquad (6.51)$$

Since each quark may have any one of the six flavours there are a large number of combinations available for the rich baryon spectrum.

The SU(3) algebra is a straightforward generalization of SU(2) to order three. Since the base vectors (6.50) have three components, the operators also in that space must be 3×3 matrices. In analogy with Eq. (6.19), any traceless Hermitian 3×3 matrix can be written as a linear combination of eight matrices λ_i which generalize the Pauli matrices. Of these, two are diagonal, corresponding to two observable properties. A relation such as Eq. (6.21) also holds. It takes the form

$$[\lambda_i, \lambda_j] = c_{ijk}\lambda_k \; , \tag{6.52}$$

where the numbers c_{ijk} are called the *structure constants* of the algebra.

We are now ready to combine the $SU(3)_c$ symmetry with the electroweak symmetry in a global symmetry for the gluonic interactions of quarks and the electroweak interactions of leptons and quarks. Group theory tells us that the most obvious way to do this is to form the direct product of the three symmetry groups,

$$SU(3)_c \otimes SU(2)_w \otimes U(1)_{B-L} \; . \tag{6.53}$$

This symmetry is referred to as the *standard model* in particle physics (not to be confused with the standard model in Big Bang cosmology). This will play an important rôle in the discussion of the primeval universe.

6.4 The Discrete Symmetries C,P,T

It is intuitively natural that the laws of physics should be independent of the choice of spacetime coordinates. Indeed all laws governing physical systems in isolation, independent of external forces, possess *translational symmetry* under the displacement of the origin in three-space as well as in four-space. Such systems also possess *rotational symmetry* in three-space. Translations and rotations are continuous transformations in the sense that a finite transformation (a translation by a finite distance or a rotation through a finite angle) can be achieved by an infinite sequence of infinitesimal transformations.

A different situation is met in the transformation from a righthanded coordinate system in three-space to a lefthanded one. This is achieved by reflecting the coordinate system in a mirror, or by replacing the x,y,z coordinates simultaneously by $-x$, $-y$, $-z$, respectively. This transformation cannot be achieved by an infinite sequence of infinitesimal transformations, and it therefore represents a *discrete transformation*. We have previously met one discrete transformation: the exchange of two identical electrons leading to exchange symmetry.

The mirror reflection in three-space is called *parity transformation*, and the corresponding *parity operator* is denoted P. Obviously every vector **v** in a righthanded coordinate system is transformed into its negative in a lefthanded coordinate system,

$$P\,\mathbf{v} = -\mathbf{v} \; . \tag{6.54}$$

This has the structure of an eigenvalue equation: **v** is an *eigenvector* of P with the eigenvalue $P = -1$. A function $f(\mathbf{r})$ of the position vector **r** is transformed by P into

$$Pf(\mathbf{r}) = f(-\mathbf{r}) \; . \tag{6.55}$$

Let us take $f(\mathbf{r})$ to be a scalar function which is either symmetric under the parity transformation, $f(-\mathbf{r}) = f(\mathbf{r})$, or antisymmetric, $f(-\mathbf{r}) = -f(\mathbf{r})$. In both cases Eq. (6.55) is an eigenvalue equation with $f(\mathbf{r})$ the *eigenfunction* of P having the eigenvalue $P = +1$ or $P = -1$, repectively. Thus scalars transform under P in two ways: those corresponding to *even parity* $P = +1$ are called (true) *scalars*, those corresponding to *odd parity* $P = -1$ are called *pseudoscalars*.

It may seem intuitively natural that the laws of physics should possess this left–right symmetry. The laws of classical mechanics in fact do, and so do Maxwell's laws of electrodynamics and Newton's and Einstein's laws of gravitation. All particles transform under P in some particular way which may be that of a scalar, a pseudoscalar, a vector or yet other. One can then consider that this is an intrinsic property, *parity* $P = \pm 1$, if the particles are eigenstates of P. The bosons are, but the fermions are not eigenstates of P because of their spinor nature (recall that the W and Z are vector bosons). However, fermion–antifermion pairs are eigenstates with odd parity, $P = -1$. The strong interactions conserve parity, and so do the electromagnetic interactions, from the evidence of Maxwell's equations. In a parity conserving universe there is no way to tell in an absolute sense which direction is left and which right.

It then came as a surprise when in 1957 the weak interactions turned out to violate left–right symmetry, in fact maximally so. In a weak interaction the intrinsic parity of a particle could change. Thus if we communicated with a being in another galaxy we could tell him in an absolute sense which direction is left by instructing him to do a β decay experiment.

A consequence of this maximal violation is the *helicity* property of the neutrinos. For a particle with spin vector **S** moving in some frame of reference with momentum **p**, the helicity is defined as

$$H = \frac{\mathbf{S} \cdot \mathbf{p}}{|\mathbf{p}|} . \tag{6.56}$$

Curiously enough, neutrinos always have their spin **S** pointing in the direction of –**p**; they are *left-handed* and their helicity is $H = -1$. If one could observe them from a frame moving faster than they do, their momentum vector would reverse its direction so that the helicity could be $H = +1$. However, the neutrinos are massless (let us for the moment ignore the possibility that they have a very small mass, see Table 4 on p. 75), so they move with the speed of light. Thus no Lorentz frame can overtake them. In consequence they remain lefthanded. Nature is not symmetric.

Analogously, antineutrinos are always right-handed. Since right-handed neutrinos and left-handed antineutrinos have never been seen, if they do exist they do not feel the weak interaction at all. In group theory language they are T_3-neutral $SU(2)_w$–singlets. We can actually generalize the lepton doublet assignments in (6.37) to include these non-observed singlets. Writing the subscripts L and R for left-handed and right-handed, respectively, the electron and antielectron families become

$$\begin{pmatrix} \nu_e \\ e^- \end{pmatrix}_L , \quad (\nu_e)_R , \quad (e^-)_R \tag{6.57}$$

$$\begin{pmatrix} e^+ \\ \bar{\nu}_e \end{pmatrix}_R , \quad (\bar{\nu}_e)_L , \quad (e^+)_L , \tag{6.58}$$

and correspondingly for the μ and τ families.

The asymmetry of the weak interactions makes the electron mainly left-handed and the positron mainly right-handed. In contrast to neutrinos they are massive, so one can make them look wrong-handed if one overtakes them in a sufficiently fast Lorentz frame. Thus the wrong-handed electron and positron singlets do exist, but their interactions are very suppressed.

The same situation occurs for the quarks. We can rewrite the first doublet in (6.45) completing it with singlets,

$$\begin{pmatrix} u \\ d \end{pmatrix}_L , \quad (u)_R , \quad (d)_R , \tag{6.59}$$

and correspondingly for the two heavier families.

Let us now introduce another discrete operator C called *charge conjugation*. The effect of C on a particle state is to turn it into its own antiparticle. For flavourless bosons like the pion this is straightforward because there is no fundamental difference between a boson and its antiboson, only the electric charge changes, e.g.

$$\text{C} \, \pi^+ = \pi^- \, . \tag{6.60}$$

Thus the charged pion is not an eigenstate of this operator, but the π^0 is. The C operator reverses the signs of all flavours, lepton numbers, and the baryon number.

For neutrinos C does not seem to be a very useful operator since it turns a left-handed neutrino into the non-observed left-handed antineutrino state,

$$\text{C} \, \nu_L = \bar{\nu}_L \, . \tag{6.61}$$

Note that the parity operator is equally useless because

$$\text{P} \, \nu_L = \nu_R \, , \tag{6.62}$$

and ν_R is also a non-observed state. However the combined operator CP is useful because it transforms left-handed neutrinos into right-handed antineutrinos, both of which are *bona fide* observed states.

$$\text{CP} \, \nu_L = \text{C}\nu_R = \bar{\nu}_R \, . \tag{6.63}$$

Weak interactions are in fact symmetric under CP to a very good approximation. Only some reactions involving kaons exhibit a tiny CP violation, of the order of 0.22% relative to CP conserving reactions. It turns out that this tiny effect is of fundamental importance for cosmology, as we shall see in the next chapter. The reason for CP violation is not yet understood.

The strong interactions violate CP with an almost equal amount of opposite sign, such that the total violation cancels, or in any case it is less than 10^{-9}. Why this is so small is also not known.

A third discrete symmetry of importance is *time reversal* T, or symmetry under inversion of the arrow of time. This is a mirror symmetry with respect to the time axis just as parity was a mirror symmetry with respect to the space axes. All physical laws of reversible processes are formulated in such a way that the replacement of time t by $-t$ has no observable effect. The particle reactions in Section 4.3 occur at the same rate in both directions of the arrow (to show this one still has to compensate for differences in phase space, i.e. the book-keeping of energy in endothermic and exothermic reactions).

Although time reversal is not very important in itself, for instance particles do not carry a conserved quantum number related to T, it is one factor in the very important combined symmetry CPT. According to our most basic notions in theoretical physics, CPT-symmetry must be absolute. It then follows from the fact that CP is not an absolute symmetry, but slightly violated, that T must be violated by an equal and opposite amount.

In a particle reaction CPT-symmetry implies that a *left-handed particle entering* the interaction region from the direction *z* is equivalent to a *right-handed antiparticle leaving* the region in the direction *z*. One consequence of this is that particles and antiparticles must have exactly the same mass, and if they are unstable, they must also have exactly the same mean life.

Needless to say, many ingenious experiments have been and still are carried out to test CPT symmetry, CP symmetry and T symmetry to ever higher precision.

6.5 Spontaneous Symmetry Breaking

As we have seen, Nature observes exactly very few symmetries. In fact, the way a symmetry is broken may be an important ingredient in the theory. As an introduction to *spontaneous breaking* of particle symmetries, let us briefly study some simple examples from other fields. For further reading, see e.g. references [1], [2], [3], [4].

Consider a cylindrical steel bar standing on one end on a solid horizontal support. A vertical downward force is applied at the other end. This system is obviously symmetrical with respect to rotations around the vertical axis of the bar. If the force is increased beyond the strength of the steel bar, it buckles in one direction or another. At that moment the cylindrical symmetry is broken.

An iron bar magnet heated above the Curie temperature 770 °C loses its magnetization. The minimum potential energy at that temperature corresponds to a completely random orientation of the magnetic moment vectors of all the electrons, so that there is no net magnetic effect. This is shown in Fig. 35 where the potential energy follows a parabola with its apex at zero magnetization. Since no magnetization direction is selected, this bar magnet possesses full rotational symmetry. The corresponding symmetry group is denoted O(3) for orthogonal rotations in three-space.

As the bar magnet cools below a temperature of 770 °C, however, this symmetry is spontaneously broken. When an external magnetic field is applied the electron magnetic moment vectors align themselves, producing a net collective macroscopic magnetization. The corresponding curve of potential energy has two deeper minima symmetrically on each side of zero magnetization, see Fig. 35. They distinguish themselves by having the north and south poles reversed. Thus the ground state of the bar magnet is in either one of these minima, not in the state of zero magnetization.

The rotational symmetry has then been replaced by the lesser symmetry of parity, or inversion of the magnetization axis. To be exact this argument actually requires that the bar magnet is infinitely long so that its moment of inertia is infinite. Then no unitary operator can rotate the north pole into the south pole. Note that the potential energy curve in Fig. 35 has the shape of a polynomial of at least fourth degree.

As a third example of a spontaneously broken symmetry we shall consider the vacuum filled with a real scalar field $\phi(x)$, where x stands for the spacetime coordinates. Recall that the electric field is a vector, it has a direction. A scalar field is like temperature, it may vary as a function of x, but it has no direction.

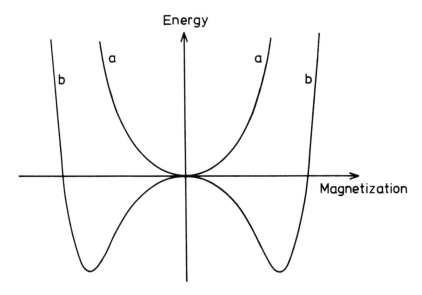

Fig. 35 Magnetization curves of bar magnet. (a) The temperature is above the Curie point 770°C and the net magnetization is zero at the potential energy minimum. (b) The temperature is below the Curie point 770°C and the net magnetization is non-vanishing at the symmetric potential energy minima

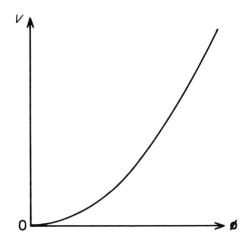

Fig. 36 Potential energy of the form (6.64) of a real scalar field ϕ

If the potential energy in the vacuum is described by the parabolic curve in Fig. 36, the total energy may be written

$$\tfrac{1}{2}(\nabla\phi)^2 + \tfrac{1}{2}m^2\phi^2 \ . \tag{6.64}$$

Here the first term is the kinetic energy which does not interest us. The second term is the potential energy $V(\phi)$ which has a minimum at $\phi = 0$ if ϕ is a classical field and m^2 is a positive number. The origin of the $\tfrac{1}{2}$ factors is of no importance to us.

If ϕ is a quantum field, it oscillates around the classical ground state $\phi = 0$ as one moves along some trajectory in spacetime. The quantum mechanical ground state is called the *vacuum expectation value* of the field. In this case it is

$$\langle \phi \rangle = 0 \ . \tag{6.65}$$

One can show that the potential (6.64) corresponds to a freely moving scalar boson of mass m (there may indeed exist such particles).

Another parameterization of a potential with a single minimum at the origin is the fourth-order polynomial

$$V(\phi) = \tfrac{1}{2}m^2\phi^2 + \frac{\lambda}{4}\phi^4 \ , \tag{6.66}$$

where λ is some positive constant.

Let us now study the potential in Fig. 37 which resembles the curve in Fig. 35 at temperatures below 770 °C. This clearly requires a polynomial of at least fourth degree. Let us use a form similar to Eq. (6.66),

$$V(\phi) = -\tfrac{1}{2}\mu^2\phi^2 + \frac{\lambda}{4}\phi^4 \ . \tag{6.67}$$

The two minima of this potential are at the field values

$$\phi_0 = \pm\mu/\sqrt{\lambda} \ . \tag{6.68}$$

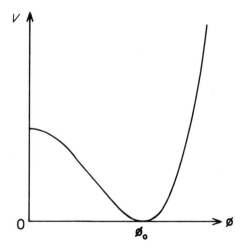

Fig. 37 Potential energy of the form (6.66) of a real scalar field ϕ

Suppose that we are moving along a spacetime trajectory from a region where the potential is given by Eq. (6.66) to a region where it is given by Eq. (6.67). As the potential changes, the original vacuum (6.65) is replaced by the vacuum expectation value $\langle \phi_0 \rangle$. Regardless of the value of ϕ at the beginning of the trajectory it will end up oscillating around ϕ_0 after a time of the order of μ^{-1}. We say that the original symmetry around the unstable *false vacuum* point at $\phi = 0$ has been broken spontaneously.

Comparing the potentials (6.67) and (6.64) we see that a physical interpretation of (6.67) would correspond to a free scalar boson with negative squared mass, $-\mu^2$! How can this be physical?

Let us replace ϕ by $\phi + \phi_0$ in the expression (6.67), then

$$V(\phi) = -\tfrac{1}{2}\mu^2(\phi + \phi_0)^2 + \frac{\lambda}{4}(\phi + \phi_0)^4 = \tfrac{1}{2}(3\lambda\phi_0^2 - \mu^2)\phi^2 + \dots . \qquad (6.69)$$

The dots indicate that terms of third and fourth order in ϕ have been left out. Comparing the coefficients of ϕ^2 in Eqs. (6.64) and (6.69) and substituting (6.68) for ϕ_0, we obtain

$$m^2 = 3\lambda\phi_0^2 - \mu^2 = 3\lambda(\mu/\sqrt{\lambda})^2 - \mu^2 = 2\mu^2 . \qquad (6.70)$$

Thus we see that the *effective mass* of the field is indeed positive so that it can be interpreted as a physical scalar boson.

The bosons in this model can be considered to move in a vacuum filled with the classical background field ϕ_0. In a way, the vacuum has been redefined, although it is just as empty as before. The only thing that has happened is that *in the process of spontaneous symmetry breaking the mass of the scalar boson has changed*. The symmetry breaking has this effect on all particles with which the scalar field interacts, fermions and vector bosons alike.

We shall now apply this model to the case of $SU(2)_w \otimes U(1)$ symmetry breaking. There are additional complications because of the group structure, but the principle is the same. The model is really the relativistic generalization of the theory of superconductivity of Ginzburg and Landau.

Let us start on a trajectory in a region of spacetime where $SU(2)_w \otimes U(1)$ is an exact symmetry. The theory requires four vector bosons as we have seen before, but in this part of the world they are massless and called B^0, W^+, W^0, W^-. We now invent such a scalar field ϕ that, as we go along the trajectory into a region where the symmetry is spontaneously broken, the vector bosons obtain their physical values.

To do this we use a trick invented by Peter Higgs. We choose the *Higgs field* ϕ to be a complex scalar $SU(2)_w$-doublet,

$$\phi = \begin{pmatrix} \phi_1 + i\phi_2 \\ \phi_3 + i\phi_4 \end{pmatrix} . \qquad (6.71)$$

The vector bosons interact with the four real components ϕ_i of the $SU(2)_w$-symmetric field ϕ. The false vacuum corresponds to the state $\phi = 0$, or

$$\phi_1 = \phi_2 = \phi_3 = \phi_4 = 0 .$$

The true vacuum which has a lower potential energy than the false vacuum corresponds to the state

$$\phi_1 = \phi_2 = 0 \,, \quad \phi_3^2 + \phi_4^2 = \; constant \; > 0 \,. \tag{6.72}$$

This potential is like the one in Fig. 37 rotated around the V axis. Thus it has rotational symmetry like a Mexican hat, see Fig. 38. All values of the potential on the circle at the bottom are equally good.

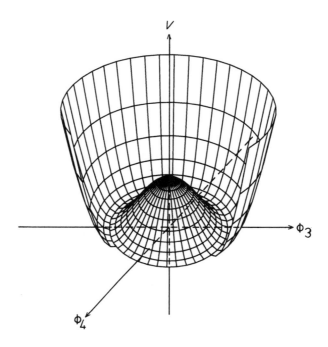

Fig. 38 'Mexican hat' potential of a complex scalar field ϕ. All the true vacuum states are located on the minimum Eq. (6.72) forming a circle at the bottom of the hat. The false vacuum is on the top of the hat at the center

If on our trajectory through spacetime we come to a point where the field has a special value like $\phi_4 = 0$, $\phi_3 > 0$ then the rotational symmetry of the Mexican hat is spontaneously broken. Just as in the case of the potential (6.69), the scalar field obtains a mass corresponding to a freely moving *Higgs boson* in a redefined vacuum. As a consequence, the vector bosons W^+, W^- interacting with the scalar field become massive. The two neutral fields B^0, W^0 form the linear combinations

$$\gamma = B^0 \cos \theta_w + W^0 \sin \theta_w \tag{6.73}$$
$$Z^0 = -B^0 \sin \theta_w + W^0 \cos \theta_w, \tag{6.74}$$

of which Z^0 becomes massive, whereas our ordinary photon γ remains massless. The reason why γ remains massless is because it is electroweakly neutral (T_3-neutral), so it does not feel the electroweak Higgs field.

Thus the Higgs boson explains the spontaneous breaking of the $SU(2)_w \otimes U(1)$ symmetry. Its mass is

$$m_\phi = 2\sqrt{\lambda} \cdot 246 \text{ GeV} . \tag{6.75}$$

Unfortunately the value of λ is unknown, so this very precise relation is useless ! At the time of writing, the Higgs boson has not yet been found, but a lower limit to its mass is about 65 GeV. But since the standard model works very well, one is confident about finding it in the next generation of particle accelerators, if not before.

Since the electroweak symmetry $SU(2)_w \otimes U(1)_{B-L}$ is the direct product of two subgroups, $U(1)_{B-L}$ and $SU(2)_w$, it depends on two coupling constants g_1, g_2 associated with the two factor groups. Their values are not determined by the symmetry, so they could in principle be quite different. This is a limitation of the electroweak symmetry. A more symmetric theory would depend on just one coupling constant g. This is one motivation for the search for a grand unified theory which would encompass the electroweak symmetry and the colour symmetry, and which would be more general than their product (6.53).

At the point of spontaneous symmetry breaking several parameters of the theory obtain specific values. It is not quite clear where the different masses of all the quarks and leptons come from, but symmetry breaking certainly plays a rôle. Another parameter is the so-called *Weinberg angle* θ_w which is related to the coupling constant of the $SU(2)_w$ subgroup. Its value is not fixed by the electroweak theory, but it is expected to be determined in whatever GUT may be valid.

6.6 Primeval Phase Transitions and GUTs

The primeval universe may have developed through phases when some symmetry was exact, followed by other phases when that symmetry was broken. The early cosmology would then be described by a sequence of *phase transitions*.

An important parameter at all times is the temperature T. When we follow the history of the Universe as a function of T, we are following a trajectory in spacetime which may be passing through regions of different vacua. In the simple model of symmetry breaking by a real scalar field ϕ having the potential (6.67) the T–dependence may be put in explicitly, as well as other dependences (denoted ...),

$$V(\phi, T, ...) = -\frac{1}{2}\mu^2\phi^2 + \frac{\lambda}{4}\phi^4 + \frac{\lambda T^2}{8}\phi^2 + ... \tag{6.76}$$

As T grows the vacuum expectation value ϕ_0 decreases so that finally the true minimum of the potential is the trivial one at $\phi = 0$. This occurs above a *critical temperature* of

$$T_c = 2\mu/\sqrt{\lambda} . \tag{6.77}$$

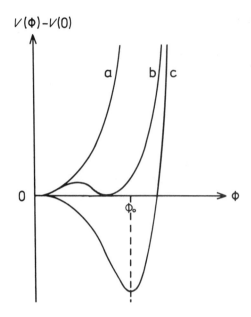

Fig. 39 Effective scalar potentials. The different curves correspond to different temperatures. At the highest temperature (a) the only minimum is at $\phi=0$, but as the temperature decreases (b) a new minimum develops spontaneously. Finally in (c), a stable minimum is reached at ϕ_0

An example of this behaviour is illustrated by the potentials in Fig. 39. The different curves correspond to different temperatures. At the highest temperature the only minimum is at $\phi = 0$, but as the temperature decreases a new minimum develops spontaneously. If there are more than one minimum, only one of them is stable. A classical example of an unstable minimum is a steam bubble in boiling water.

Let us now construct a possible scenario for the early history of the Universe. We start at the same point as we did in Section 4.3, about one microsecond after Big Bang when the energy of the particles in thermal equilibrium was about 1 GeV, but now we let time run backwards. For the different scales we can refer to Fig. 32 and the colour plate 1, Early History of the Universe at the beginning of the book.

E ≈ 1 GeV

At this time the quarks are confined into protons and neutrons. The colour symmetry $SU(3)_c$ is still valid, but it is in no way apparent because the hadrons are colour-neutral singlets. The colour force mediated by gluons is also not apparent: a vestige of it remains in the form of strong interactions between hadrons. It appears that this force is mediated by mesons, themselves quark bound states.

There is no trace of the weak isospin symmetry $SU(2)_w$, so the weak and electromagnetic interactions look quite different. Their strengths are very different, and the masses of the leptons are very different. Only the electromagnetic gauge symmetry U(1) is exactly valid, as is testified to by the conservation of electric charge.

1 GeV $\overset{<}{\sim}$ E $\overset{<}{\sim}$ 100 GeV

As the temperature increases through this range, the electroweak identity of the weak and electromagnetic interactions becomes progressively clearer. The particles contributing to the effective degrees of freedom in the thermal soup are quarks, gluons, leptons and photons. All the fermions are massive, except perhaps the neutrinos.

The electroweak symmetry $SU(2)_w \otimes U(1)_{B-L}$ is broken, as is testified to by the very different quark masses and lepton masses. The electroweak force is mediated by massless photons and virtual $W^{\pm}, Z^0$ vector bosons. The latter do not occur as free particles, because the energy is still lower than their rest masses near the upper limit of this range. The $SU(3)_c \otimes U(1)$ symmetry is of course exact, and the interactions of quarks are ruled by the colour force.

E $\approx$ 100 GeV.

This is about the rest mass of the W and Z, so they freeze out of thermal equilibrium. The Higgs boson also freezes out now, if it has not done so already at higher temperature. Our ignorance here depends on the lack of experimental information about its mass.

There is no difference between weak and electromagnetic interactions: there are charged-current electroweak interactions mediated by the $W^{\pm}$ and neutral-current interactions mediated by the Z^0 and γ. However, the electroweak symmetry is imperfect because of all the different masses.

E $\approx$ 1 TeV.

Up to this energy our model of the Universe is fairly reliable, because this is the limit of present-day laboratory experimentation. Here we encounter the phase transition between exact and spontaneously broken $SU(2)_w \otimes U(1)_{B-L}$ symmetry. The end of electroweak unification is marked by the massification of the vector boson fields, the scalar Higgs field, and the fermion fields.

1 TeV $\overset{<}{\sim}$ E $\overset{<}{\sim}$ 10^{14-15} GeV

Let us introduce an abbreviation G_s for the 'standard' symmetry group (6.53) which is exactly valid in this range,

$$G_s = SU(3)_c \otimes SU(2)_w \otimes U(1)_{B-L} . \qquad (6.78)$$

As we have seen, laboratory physics has led us to construct this theory, so it is fairly well understood although experimental information above 1 TeV is lacking. The big question is what new physics will appear in this enormous energy range. The possibility that nothing new appears is called 'the desert'.

The new physics could be a higher symmetry which would be broken at the lower end of this energy range. Somewhere there would then be a phase transition between the exactly symmetric phase and the spontaneously broken phase. Even in the case of a 'desert' one expects a phase transition to a *grand unified theory* (GUT) at 10^{14} or 10^{15} GeV.

If the GUT symmetry G_{GUT} breaks down to G_s through intermediate steps, the phenomenology could be very rich. For instance, there are *subconstituent models* building leptons and quarks out of elementary particles of one level deeper elementarity. These subconstituents would freeze out at some intermediate energy, condensing into lepton and quark bound states. The forces binding them are in some models called *technicolour forces*.

10^{15} GeV $\overset{<}{\sim}$ E $\overset{<}{\sim}$ 10^{19} GeV

Let us devote some time to GUTs which might be exact in this range. The unification of forces is not achieved very satisfactorily within the G_s symmetry. It still is the direct product of three groups, thus there are in principle three independent coupling constants g_1, g_2, g_3 associated with it. Full unification of the electroweak and colour forces would imply a symmetry group relating those coupling constants to only one g. The specific values of the coupling constants are determined—by accident?—at the moment of spontaneous G_{GUT} breaking.

Below the energy of electroweak unification we have seen that the electromagnetic and weak coupling strengths are quite different. As the energy increases, their relative strengths change. Thus the coupling constants are functions of energy; one says that they are *running*. If one extrapolates the running coupling constants from their known low-energy regime, they almost intersect in one point, see Fig. 40 [5]. The energy of that point is between 10^{13} and 10^{15} GeV, so if there is a GUT, that must be its unification scale and the scale of the masses of its vector bosons and Higgs bosons.

The fact that the coupling strengths do not run together to exactly one point is actually a quite interesting piece of information. It could imply that there exist intermediate symmetries in the 'desert' between G_{GUT} and G_s. Their effect on the lines in Fig. 40 would be to change their slopes at the intermediate energy. The present extrapolation to GUT energy would then be wrong, and the lines could after all meet exactly at one point.

As I have pointed out several times, it is not understood why the leptons and the quarks come in three families. The symmetry group requires only one family, but if Nature provides us with more, why precisely three? This is one incentive to search for larger unifying symmetries. However, *family unification theories* (FUT) need not be the same as GUT.

The standard model leaves a number of other questions open which one would very much like to have answered within a GUT. There are too many free parameters in G_s. Why are the electric charges such as they are? How many Higgs scalars are there? What is the reason for CP-violation? The only hint is that CP-violation seems to require (but not explain) at least three families of quarks.

Why is parity maximally violated? Could it be that the left–right asymmetry is only a low-energy artefact which disappears at higher energies? There are left–right symmetric models containing G_s, in which the righthanded particles interact with their own righthanded vector bosons, which are one or two orders of magnitude heavier than their lefthanded partners. Such models then have intermediate unification scales between the G_s and G_{GUT} scales.

In a GUT all the leptons (6.57), (6.58) and quarks (6.59) should be components of the same field. In consequence there must exist *leptoquark* vector bosons X which can

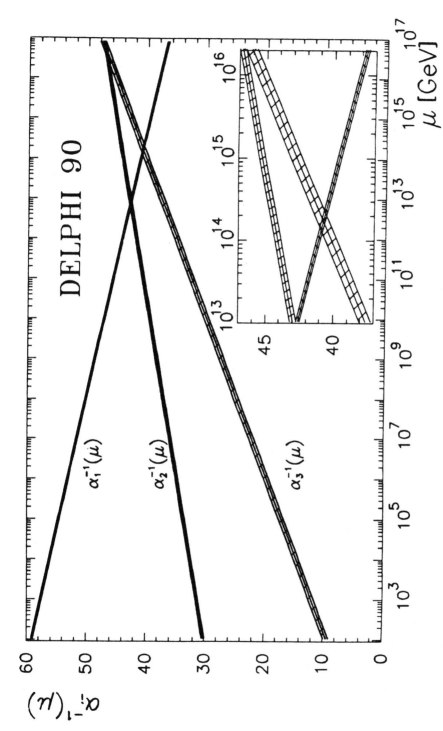

Fig. 40 Evolution of the inverse of the coupling constants α_1, α_2, α_3 for the symmetry groups $U(1)_{B-L}$, $SU(2)_w$, and $SU(3)_c$, respectively, using M_Z from DELPHI data. The width of the bands indicate the experimental precision in 1991. The energy is denoted μ. Reproduced from reference [5] by permission of Elsevier Science Publishers BV

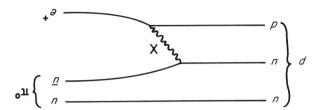

Fig. 41 Proton decay Feynman diagram

transform a quark into a lepton in the same way as the colour(i)–anticolour(j) gluons can transform the colour of quarks. This has important consequences for the stability of matter: the quarks in a proton could decay into leptons, for instance as depicted in Fig. 41, thereby making protons unstable.

Experimentally we know that the mean life of protons must exceed the age of the Universe by many orders of magnitude. Sensitive underground detectors have been waiting for years to see a single proton decay in large volumes of water. There are about 10^{33} protons in a detector of 2000 tons of water, so if one would see one decay in a year the proton mean life would be

$$\tau_p \gtrsim 10^{33} \text{years}. \tag{6.79}$$

This is about the present experimental limit. It sets stringent limits on the possible GUT candidates, and it has already excluded a GUT based on the symmetry group SU(5), which offered the simplest scheme of quark–lepton cohabitation in the same multiplets.

One noteworthy remaining GUT candidate is *supersymmetry* which includes a large number of new particles, some of which should be seen at energies already available. In this theory there is a conserved multiplicative quantum number, *R parity*, defined by

$$R = (-1)^{3B+L+2s}, \tag{6.80}$$

where B, L and s are baryon number, lepton number and spin, respectively. All known particles have $R = +1$, but the theory allows an equal number of supersymmetric partners, *sparticles*, having $R = -1$. Conservation of R ensures that supersymmetric particles can only be produced pairwise, as sparticle–antisparticle pairs. The lightest sparticles must therefore be stable, just as the lightest particles.

Although all GUTs are designed to answer some of the questions and relate some of the parameters in the standard model, they still have the drawback of introducing large numbers of new particles, vector bosons and Higgs scalars, all of which are yet to be discovered.

E $\gtrsim 10^{19}$ GeV.

We have now reached the *Planck mass* defined in Eq. (3.13) as the energy scale where gravitational and quantum effects are of equal importance, so we can no longer do particle physics neglecting gravitation. Unfortunately there is as yet no theory including quantum mechanics and gravitation. Thus we are forced to stop here at the *Planck time*, a 'mere' 10^{-43} s after Big Bang, because of a lack of theoretical tools. But we shall come back to these earliest times in connection with models of cosmic inflation and in the final chapter.

Problems

1. Are the raising and lowering operators S_+ and S_- unitary or Hermitian ?
2. Verify the commutation relations (6.21) and anticommutation relations (6.22) for the Pauli spin matrices (6.20).
3. Derive Eqs. (6.29) and (6.30). Hint: Define a lowering operator $S_- = S_-^a + S_-^b$ for the two-electron spin system, and operate with this on the $|1,1\rangle$ state (6.27). Equation (6.30) is found as the orthogonal linear combination.
4. Write down the weak hypercharge operator Y (6.42) in matrix form.
5. The s quark is assigned the value of strangeness $S = -1$. Generalize the relation (6.34) to include strangeness so that it holds true for the K mesons defined in Eqs. (6.44). Note that of all the quark–antiquark systems possible with three quarks, only five are listed in Eqs. (6.44). What are the properties of the missing ones ?
6. Show by referring to the quark structure that the K mesons are not eigenstates of the C operator.
7. All known baryons are qqq-systems. Use the u,d,s quarks to compose the 27 ground state baryons, and derive their charge and strangeness properties. Plot these states in (I_3,Y)-space.
8. In SU(3) the Pauli spin matrices generalize to eight 3×3 matrices which have the commutation relations (6.52). Derive them !
9. Derive the expression (6.51).
10. Is the parity operator P defined in Eq. (6.55) Hermitian ?

References

1. M. Chaichian and N. F. Nelipa, *Introduction to Gauge Field Theories*, Springer-Verlag, Berlin, 1984.
2. E. W. Kolb and M. S. Turner, *The Early Universe*, Addison-Wesley Publ. Co., Reading, Mass, 1990.
3. P. D. B. Collins, A. D. Martin and E.J. Squires, *Particle Physics and Cosmology*, John Wiley & Sons, New York, 1989.
4. A. Linde, *Particle Physics and Inflationary Cosmology*, Harwood Academic Publishers, London, 1990.
5. U. Amaldi, W. de Boer and H. Fürstenau *Physics Letters B*, **260** (1991) 447.

7 Problems with the Big Bang model

The standard Big Bang model was defined as an adiabatically expanding universe described by a Friedman–Robertson–Walker (FRW) metric having an initial singularity. This model has, as so far presented, been essentially a success story. We are now going to correct that optimistic picture. The model is in fact afflicted with serious difficulties and open problems, some of which we shall describe in this chapter.

Baryons were produced from quarks in a process we may call *baryosynthesis*. This process is not well understood. Indeed, the number of baryons stands in a ratio to the number of photons which is very small, yet the conclusion from the FRW model is that it should be nine orders of magnitude smaller! Equally surprising is the absence of antibaryons. These problems are discussed in Section 7.1 and some plausible solutions are outlined.

In Section 7.2 we collect several problems which all have something to do with the structure of spacetime: the horizon problem, topological problems of which the monopole problem is an example, and the curvature or flatness problem. The solutions to all of these will require rather drastic modifications of the standard model to which we shall come in the next chapter.

In Section 7.3 we shall describe the dynamical evidence that a large fraction of the gravitating mass in the Universe is non-luminous and composed of some unknown kind of non-baryonic 'dark' matter. We leave the discussion of particle candidates and the relation between dark matter and galaxy formation to Chapter 9.

7.1 Baryosynthesis

In Section 5.3 we noted that the ratio η of the baryon number density N_B to the photon number density N_γ is very small. We arrived in Eq. (5.46) at the value

$$2.6 \overset{<}{\sim} 10^{10} \, \eta \overset{<}{\sim} 3.3 \tag{7.1}$$

from nucleosynthesis evidence.

Before the baryons were formed (at about 200 MeV) the conserved baryon number B was carried by quarks. Thus the total value of B carried by all protons and neutrons today should equal the total value carried by all quarks. It is very surprising then that η should be so small today, because when the quarks, leptons and photons were in thermal equilibrium there should have been equal numbers of quarks and antiquarks, leptons and antileptons, and the value of B was equal to the total leptonic number L,

$$N_B = N_{\bar{B}} = N_L \approx N_\gamma \tag{7.2}$$

because of the $U(1)_{B-L}$ symmetry.

The other surprising thing is that no antibaryons seem to have survived. At temperatures above 1 GeV quarks and antiquarks were in thermal equilibrium with the photons because of reactions such as (4.31–4.34), as well as

$$\gamma + \gamma \leftrightarrow q + \bar{q} \; . \tag{7.3}$$

These reactions conserve baryon number so that every quark produced or annihilated is accompanied by one antiquark produced or annihilated. Thus it would be reasonable to expect the number densities of baryons and antibaryons to be equal today.

We are then faced with two big questions. What caused the smallness of η ? And why does the Universe not contain antimatter (antiquarks and positrons).

When the baryons and antibaryons become non-relativistic the numbers of baryons and antibaryons were reduced by annihilation, so N_B decreased rapidly by the exponential factor in the Maxwell–Boltzmann distribution (4.42). The number density of photons N_γ is given by Eq. (4.5) at all temperatures. Thus the temperature dependence of η is

$$\eta = \frac{N_B}{N_\gamma} = \frac{\sqrt{2\pi}}{4.808} \left(\frac{m_N}{kT} \right)^{3/2} e^{-m_N/kT} \; . \tag{7.4}$$

When the annihilation rate became slower than the expansion rate the value of η was frozen, and thus comparable to its value today (7.1). The freeze-out occurs at about 20 MeV when η has reached the value

$$\eta \simeq 6.8 \times 10^{-19} \; . \tag{7.5}$$

But this is a factor 5×10^8 too small ! Thus something must be seriously wrong with our initial condition (7.2).

Let us turn to the question of the absence of antimatter. We know that the Earth is matter and since the solar wind does not produce annihilations with the earth, or with other planets, we know that the solar system is matter. Since no gamma rays are produced in the interaction of the solar wind with the local interstellar medium, we know the interstellar medium, and hence our Galaxy, is matter. The main evidence that other galaxies are not composed of antimatter comes from cosmic rays. Our Galaxy contains free protons (cosmic rays) with a known velocity spectrum very near the speed of light. A fraction of these particles have sufficiently high velocities to escape from the Galaxy. These protons would annihilate with antimatter cosmic rays in the intergalactic medium or in collisions with antimatter galaxies if they existed and produce characteristic gamma-rays many orders of magnitude more frequently than what has been seen,

$$\frac{N_{\bar{B}}}{N_B} \simeq 10^{-4} \ .$$

This small number essentially rules out the possibility that other galaxies' cosmic rays are composed of antimatter. There are many other pieces of evidence against antimatter, but the above arguments are the strongest.

Thus the only reasonable conclusion is that $N_{\bar{B}}$ and N_B must have started out slightly different while they were in thermal equilibrium by the amount

$$N_B - N_{\bar{B}} \simeq \eta N_\gamma \ . \tag{7.6}$$

Subsequently all antibaryons were annihilated, and the small excess ηN_γ of baryons is what remained. This idea is fine, but the basic problem has not been removed; we have only pushed it further back to earlier times. We still require an explanation for the primordial $B\bar{B}$-asymmetry.

The solution may come from some early phase transition. Let us consider theories in which a $B\bar{B}$-asymmetry could arise. For this three conditions must be met.

First, the theory must contain reactions violating baryon number conservation. Grand unified theories are obvious candidates for a reason we have already met in Section 6.6. We noted there that GUTs are symmetric with respect to leptons and quarks, because they are components of the same field and GUT forces do not see any difference. In consequence, GUTs contain leptoquarks X,Y which transform quarks into leptons. Reactions involving X,Y do explicitly violate baryon number conservation since the quarks have $B = \frac{1}{3}$, but leptons are neutral, $B = 0$. The baryon number then changes by an amount ΔB, as for instance in the decay reactions

$$X \rightarrow e^- + d \ , \quad \Delta B = +\tfrac{1}{3} \tag{7.7}$$

$$X \rightarrow \bar{u} + \bar{u} \ , \quad \Delta B = -\tfrac{2}{3} \ . \tag{7.8}$$

Secondly, there must be C and CP violation in the theory, as these operators change particles into antiparticles. If the theory were C and CP symmetric, even the baryon-violating reactions (7.7) and (7.8) would be matched by equally frequently occurring reactions with opposite ΔB, so no net $B\bar{B}$-asymmetry would result. In fact, we want baryon production to be slightly more frequent than antibaryon production.

Thirdly, we must require these processes to occur out of thermal equilibrium. In thermal equilibrium there is no net production of baryon number because the reactions (7.7) and (7.8) go as frequently in the opposite direction. Hence the propitious moment is the phase transition when the X-bosons are freezing out of thermal equilibrium and decay. If we consult the timetable in Section 6.6 this would happen at about 10^{14} GeV, the moment for the phase transition from the GUT symmetry to its spontaneously broken remainder.

The scenario is therefore the following. At some energy $E_X = kT_X$ which is of the order of the rest masses of the leptoquark bosons X,

$$E_X \simeq M_X c^2 , \tag{7.9}$$

all the X,Y vector bosons, the Higgs bosons, the W,B vector bosons of Eqs. (6.73–6.74) and the gluons are in thermal equilibrium with the leptons and quarks. The number density of each particle species is about the same as the photon number density and the relations (7.2) hold.

When the age of the Universe is still young, as measured in Hubble time τ_H, compared to the mean life $\tau_X = \Gamma_X^{-1}$ of the X bosons, there are no X decays and therefore no net baryon production. The X bosons start to decay when

$$\Gamma_X \gtrsim \frac{1}{\tau_H} = H . \tag{7.10}$$

This is just like the condition (4.62) for the decoupling of neutrinos. The decay rate Γ_X is proportional to the mass M_X,

$$\Gamma_X = \alpha M_X , \tag{7.11}$$

where α is essentially the coupling strength of the GUT interaction. It depends on the details of the GUT and the properties of the X boson.

We next make use of the expression (4.60) for the temperature dependence of the expansion rate H. Let us replace there the Newtonian constant G by its expression in terms of the Planck mass M_P Eq. (3.13). We then find

$$H = \sqrt{\frac{16\pi \hbar a}{3c} g_{eff}(T)} \frac{T^2}{M_P} . \tag{7.12}$$

Substituting this H and the expression (7.11) into the condition (7.10) for thermal equilibrium we obtain

$$A M_X \gtrsim \sqrt{g_{eff}(T)} \frac{T^2}{M_P} , \tag{7.13}$$

where all the constants have been lumped into A. Solving for the temperature squared, we find

$$T^2 \lesssim \frac{A M_X M_P}{\sqrt{g_{eff}(T)}} . \tag{7.14}$$

At the temperature T_X the effective degrees of freedom g_{eff} are approximately 100. The condition (7.14) then gives an upper limit to the X boson mass,

$$M_X \lesssim A' \frac{M_P}{\sqrt{g_{eff}(T)}} = A' \frac{1.2 \times 10^{19} \text{ GeV}}{\sqrt{100}} \simeq A' \times 10^{18} \text{ GeV} , \tag{7.15}$$

where A' includes all constants left out.

Thus if the mass M_X is heavier than $A' \times 10^{18}$ GeV the X bosons are stable at energies above M_X. Let us assume that this is the case. As the energy drops below M_X the X and $\bar{X}$ bosons start to decay, producing the net baryon number required. The interactions must be such that the decays really take place out of equilibrium, that is, the temperature of decoupling should be above M_X. Typically bosons decouple at about $M_X/20$, so it is not trivial to satisfy this requirement.

Let us now see how C and CP violation can be invoked to produce a net $B\bar{B}$-asymmetry in X and $\bar{X}$ decays. We can limit ourselves to the case when the only decay channels are (7.7) and (7.8), and correspondingly for the $\bar{X}$ channels. For these channels we tabulate in Table 7 the net baryon number change ΔB and the branching fractions $\mathrm{BR}_i = \Gamma(X \rightarrow \text{channel } i)/\Gamma(X \rightarrow \text{all channels})$ in terms of two unknown parameters r and $\bar{r}$.

Table 7

i	Channel i	ΔB_i	BR_i
1	$X \rightarrow \bar{u} + \bar{u}$	$-\frac{2}{3}$	r
2	$X \rightarrow e^- + d$	$+\frac{1}{3}$	$1-r$
3	$\bar{X} \rightarrow u + u$	$+\frac{2}{3}$	$\bar{r}$
4	$\bar{X} \rightarrow e^+ + \bar{d}$	$-\frac{1}{3}$	$1-\bar{r}$

The baryon number produced in the decay of one pair of $X, \bar{X}$ vector bosons weighted by the different branching ratios is then

$$\Delta B = r\Delta B_1 + (1-r)\Delta B_2 + \bar{r}\Delta B_3 + (1-\bar{r})\Delta B_4 = \bar{r} - r \ . \tag{7.16}$$

If C and CP symmetry are violated, r and $\bar{r}$ are different, and we obtain the desired result $\Delta B \neq 0$.

Suppose that the number density of X and $\bar{X}$ bosons is N_X. We now want to generate a net baryon number density

$$N_B = \Delta B \ N_X \simeq \Delta B \ N_\gamma$$

by the time the Universe has cooled through the phase transition at T_{GUT}. After that the baryon number is absolutely conserved and the further decrease in N_B only follows the expansion. However, the photons are bosons, so their absolute number is not conserved. It then follows that the value of η may be changing somewhat. Thus if we want to confront the baryon production ΔB required at T_{GUT} with the present-day value of η a more useful quantity is the baryon number per unit entropy N_B/s. Recall that the entropy density of photons is

$$\frac{s}{V} = 1.80 \ g_{eff}(T) \ N_\gamma \tag{7.17}$$

from Eq. (4.75). At temperature T_{GUT} the effective degrees of freedom is about 100, so the baryon number per unit entropy is

$$\frac{N_B}{s} = \frac{\Delta B}{1.80 \, g_{eff}(T_{GUT})} \simeq \frac{\Delta B}{180} \ . \tag{7.18}$$

Clearly this ratio scales with $g_{eff}(T)$. Thus to observe a present-day value of η in the range (7.1) the GUT should be chosen such that it yields

$$\Delta B = \frac{g_{eff}(T_{GUT})}{g_{eff}(T_0)} \, \eta \simeq \frac{100}{3.36} \, \eta = (0.8 - 1.0) \, 10^{-8} \ . \tag{7.19}$$

This is within the possibilities of various GUTs.

One may of course object that this solution of the baryosynthesis problem is only speculative since it rests on the assumption that Nature exhibits a suitable GUT symmetry, for which we have no other indications. However, this solution is not restricted to a phase transition at GUT time. The three conditions referred to could perhaps be met at some later phase transition. This is precisely the conclusion one is forced to by inflationary models (in Chapter 8), because inflation turns out to wash out any baryon asymmetry generated at GUT time. At this point we must content ourselves with the conclusion that the baryosynthesis problem does not force us to abandon the Big Bang model.

7.2 Spacetime Problems

In this section we shall discuss three problems related to properties of the spacetime of the causally connected Universe: the *horizon problem* associated to its size at different epochs, the *monopole problem* associated with possible topological defects, and the *flatness problem* associated with its metric.

It is an unavoidable consequence of the finite age of the Universe that the particle horizon today is larger than at any earlier time. This implies that the Universe is larger than that which our past light cone encloses today, so that with time we will become causally connected with new regions as they move in across our horizon. This renders the question of the full size of the whole Universe meaningless, the only meaningful size being the diameter of its horizon at a given time.

Recall that we calculated in Eq. (5.15) that the particle horizon at the last scattering surface subtends today an angle of only $1°$ on our present horizon. This was illustrated in Fig. 28. We noted then that the existence of cosmic structures exceeding $1°$ in size implied that these structures must be of a much more recent date than the last scattering. The age of the Universe at temperature 20 MeV was $t_U = 2$ ms and the distance scale $2ct_U$. The amount of matter inside that horizon was only about $10^{-5} \, M_{\odot}$, which is very far from what we see today: matter is separated into galaxies of mass $10^{12} \, M_{\odot}$. The size of present superclusters is so large that their mass must have been assembled from vast regions of the Universe which were outside the particle horizon at $t_U = 2$ ms. But then they must have been formed quite recently, in contradiction with the age of the quasars and galaxies they contain.

Even more serious problems emerge as we approach very early times. At GUT time the temperature of the cosmic background radiation was $T_{GUT} \simeq 1.2 \times 10^{28}$ K, or a factor

$$\frac{T_{GUT}}{T_0} \simeq 4.4 \times 10^{27}$$

more than now. This is the factor by which the linear scale S(t) has increased since the time t_{GUT}. If we take the present Universe to be of size 2000 h^{-1} Mpc = 6×10^{25} m, its linear size was only 2 cm at GUT time.

Note, however, that linear size and horizon are two different things. The horizon size depends on the time perspective back to some earlier time. Thus the particle horizon today has increased since t_{GUT} by almost the square of the linear scale factor, or by

$$\frac{t_0}{t_{GUT}} = \left(\frac{g_{\mathit{eff}}(T_{GUT})}{g_{\mathit{eff}}(T_0)} \right)^{\frac{1}{2}} \left(\frac{T_{GUT}}{T_{LSS}} \right)^2 \left(\frac{T_{LSS}}{T_0} \right)^{\frac{3}{2}} \simeq 3.1 \times 10^{54} . \tag{7.20}$$

At GUT time the particle horizon was only 3.8×10^{-27} cm. It follows that to arrive at the present homogeneous Universe, the homogeneity at GUT time must have extended out to a distance 5×10^{26} times greater than the distance of causal contact! Why did the GUT phase transition happen simultaneously in a vast number of causally disconnected regions? Concerning even earlier times one may ask the same question about the Big Bang. Obviously this poses a serious problem to the standard Big Bang model. We shall return to this in Section 8.1.

In all regions where the GUT phase transition was completed, several important parameters such as the coupling constants, the charge of the electron, the masses of the vector bosons and Higgs bosons obtained values which were going to characterize the present Universe. (Recall that the coupling constants are functions of energy, as was illustrated in Fig. 40, and the same is true for particle masses.) One may wonder why they obtained the same value in all causally disconnected regions.

The Higgs field had to take the same value everywhere, because this is uniquely dictated by what is its ground state. But one might expect that there would be domains where the phase transition was not completed, so that certain remnant symmetries froze in. The Higgs field could then settle to different values, causing some parameter values to be different. The physics in these domains would then be different, and so the domains would have to be separated by *domain walls* which are *topological defects* of spacetime. Such domain walls would contain enormous amounts of energy and in isolation they would be indestructible. Intersecting domain walls would produce other types of topological defects such as *loops* or *cosmic strings* wiggling their way through the Universe. No evidence for topological defects has been found, perhaps fortunately for us, but they may still lurk outside our horizon.

A particular kind of topological defect is a *magnetic monopole*. Ordinarily we do not expect to be able to separate the north and south poles of a bar magnet into two independent particles. As is well known, cutting a bar magnet in two produces two dipole bar magnets. Maxwell's equations account for this by treating electricity and magnetism differently: there is an electric source term containing the charge e, but there is no magnetic source term. Thus free electric charges exist, but not free magnetic charges. Stellar bodies may have large magnetic fields, but no electric fields.

Paul A.M. Dirac (1902–1984) suggested in 1931 that the quantization of the electron charge was the consequence of the existence of at least one free magnetic monopole with magnetic charge

$$g_M = \frac{1}{2} \frac{\hbar\, cn}{e} \simeq 68.5\, en \,, \qquad (7.21)$$

where e is the charge of the electron and n is an unspecified integer. This would then modify Maxwell's equations, rendering them symmetric with respect to electric and magnetic source terms. Free magnetic monopoles would have drastic consequences, for instance destroying stellar magnetic fields.

In the GUT phase transition there arises at least one monopole per horizon volume. The monopole number density at GUT time was then

$$N_M(t_{GUT}) = (3.8 \times 10^{-27}\ \mathrm{cm})^{-3} \,,$$

and the linear scale has grown by a factor 3.1×10^{27}. Nothing could have destroyed them except monopole–antimonopole annihilation, so the monopole density today should be

$$N_M(t_0) \simeq (3.1 \times 3.8\ \mathrm{cm})^{-3} \simeq 6.1 \times 10^{-4}\ \mathrm{cm}^{-3} \,. \qquad (7.22)$$

This is quite a substantial number compared with the proton density which is at most 1.7×10^{-7} per cm^3. Monopoles circulating in the galaxy would take their energy from the galactic magnetic field. Since the field survives, this sets a very low limit to the monopoles flux called the *Parker bound*. Experimental searches for monopoles have not yet become sensitive enough to test the Parker bound, but they are certainly in gross conflict with the above value of N_M, the present experimental upper limit to N_M is 25 orders of magnitude smaller than N_γ.

Monopoles are expected to be superheavy,

$$m_M \gtrsim \frac{m_X}{\alpha_{GUT}} \simeq 10^{16}\ \mathrm{GeV} \simeq 0.02\mu\mathrm{g} \,. \qquad (7.23)$$

Combining this mass with the number densities (5.24) and (7.22) the density parameter of monopoles becomes

$$\Omega_M = \frac{N_M}{N_B} \frac{m_M}{m_B} \Omega_B \simeq 6 \times 10^{17} \,. \qquad (7.24)$$

This is in flagrant conflict with the upper limit of the density parameter of matter of any kind, $\Omega \leq 1.5$, derived in Eq. (3.65). Such a universe would be closed and its maximal lifetime would be only a fraction of the age of the present Universe, of the order of

$$t_{max} = \frac{\pi}{2H_0\sqrt{\Omega_M}} \simeq 13\ \mathrm{yr} \,. \qquad (7.25)$$

Monopoles have other curious properties as well. Unlike the leptons and quarks which appear to be pointlike down to the smallest distances measured (10^{-19} m) the monopoles have an internal structure. All their mass is concentrated within a core of about 10^{-30} m, with the consequence that the temperature in the core is of GUT scale or more. Outside that core there is a layer populated by the X leptoquark vector bosons, and outside that at about 10^{-17} m there is a shell of W and Z bosons. This structure may affect the stability of matter, as we shall see in Section 10.2.

The third problem associated with the structure of spacetime is called the *flatness problem* or the *curvature problem*. Recall that in a flat universe the curvature parameter k vanishes and the density parameter is $\Omega = 1$. This is obvious from Eq. (3.66) where k and Ω are related by

$$\Omega - 1 = \frac{kc^2}{\dot{S}^2} .$$

During the radiation era the energy density ε_r or ρc^2 in Friedman's equation (3.20) is proportional to S^{-4}. It then follows that

$$\dot{S}^2 \propto S^{-2}. \tag{7.26}$$

To be specific, consider GUT time when the linear scale is some 10^{27} times smaller than today; then

$$\Omega - 1 \propto S^2 \simeq 10^{-54} . \tag{7.27}$$

Thus the Universe at that time must have been flat to within the 54th decimal place, a totally incredible situation. If this were not so the Universe would either have reached its maximum size within one Planck time (10^{-43} s), and thereafter collapsed into a singularity, or it would have dispersed into a vanishingly small energy density. The only natural values for Ω are therefore 0, 1 or infinity, whereas to generate a universe surviving several Gyr requires an incredible finetuning. It would also be an accident of timing if Ω_0 happened to be about 0.1 today, whereas it was almost exactly 1 during all previous time, until very recently.

Another way to state this problem is to consider the dimensionless quantity

$$d = (\Omega_0 - 1)\frac{H_0^2 \hbar^2}{E_r^2} , \tag{7.28}$$

where E_r is the energy of radiation. This is constructed to be independent of S since $E_r \propto S^{-1}$. Inserting the present value of E_r and the upper limits of Ω and H_0, we find the value

$$d \stackrel{<}{\sim} 0.4 \times 10^{-58} . \tag{7.29}$$

It is quite surprising that such small numbers occur in Nature. It would seem much more natural if d were of the order of unity or exactly zero. Thus there must be some deep reason not explained by the Big Bang model.

7.3 Dynamical Dark Matter

Let us consider in turn dynamical systems on different scales: galaxies, small galaxy groups and rich galaxy clusters.

The spiral galaxies are stable gravitationally bound systems in which matter is composed of stars and interstellar gas. Most of the observable matter is in a relatively thin disc, where stars and gas travel around the galactic centre on nearly circular orbits. By observing the Doppler shift of the integrated starlight and the radiation at $\lambda = 21$ cm from the interstellar hydrogen gas one finds that galaxies rotate. If the circular velocity at radius R is v in a galaxy of mass M, the condition for stability is that the centrifugal acceleration should equal the gravitational pull

$$\frac{v^2}{R} = \frac{GM}{R^2} \ .$$

(7.30)

In other words, the radial dependence of the velocity of matter rotating in a disc is expected to follow Kepler's law

$$v = \sqrt{\frac{GM}{R}} \ .$$

(7.31)

Visible starlight traces velocity out to radial distances typically of the order of 10 kpc, and interstellar gas out to 20–50 kpc. The surprising result from measurements of galaxy rotation curves is that the velocity does not follow the $1/\sqrt{R}$ law (7.31), but stays constant after a maximum at about 5 kpc, see Fig. 42. Assuming that the disc surface brightness is proportional to the surface density of luminous matter, one derives a circular speed which is typically more than a factor 3 lower than the speed of the outermost measured points (see, for example, reference [1]). This implies that the calculated gravitational field is too small by a factor of 10 to account for the observed rotation.

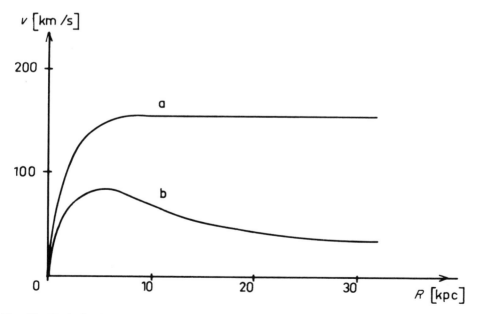

Fig. 42 Typical galaxy rotation curves, (a) derived from the observed Doppler shift of the 21 cm line of atomic hydrogen relative to the mean; (b) prediction from the radial light distribution

There are only a few possible solutions to this problem. One is that the theory of gravitation is wrong. It is possible to modify *ad hoc* Kepler's inverse square law or Newton's assumption that G is a constant, but the corresponding modifications cannot be carried out in the relativistic theory. There are also interesting attempts to generalize the Einstein tensor by including terms containing higher derivatives of the curvature tensor.

Another possibility is that spiral galaxies have magnetic fields extending out to regions of tens of kiloparsecs where the interstellar gas density is low and the gas dynamics may easily be modified by such fields [2]. But this argument works only on the gas halo, and does not affect the velocity distribution of stars. Also, the existence of magnetic fields of sufficient strength remains to be demonstrated; in our Galaxy it is only a few microgauss. For the time being this idea is this only a theoretical one.

The solution attracting most attention is that there exist vast amounts of non-luminous *dark matter* beyond what is accounted for by stars and hydrogen clouds. One natural place to look for dark matter is in the neighbourhood of the solar system. Already in 1922 Jacobus C. Kapteyn deduced that the total density in the local neighbourhood is about twice as large as the luminous density in visible stars. Although the result is somewhat dependent on how large one chooses this 'neighbourhood', modern dynamical estimates are similar.

The luminous parts of galaxies, as evidenced by radiation of baryonic matter in the visible, infrared and X-ray spectra, account only for

$$\Omega_{lum} < 0.01 \tag{7.32}$$

of the total density parameter. The internal dynamics implies that the galaxies are embedded in extensive halos of dark matter, of the order of

$$\Omega_{halo} > 0.03 - 0.10 \ . \tag{7.33}$$

In fact, to explain the observations, the radial mass distribution $M(R)$ must be proportional to R,

$$v = \sqrt{\frac{GM(R)}{R}} \ \propto \ \sqrt{\frac{GR}{R}} = constant. \tag{7.34}$$

The radial density distribution is then

$$\rho(R) \propto R^{-2} \ . \tag{7.35}$$

This is precisely the distribution one would obtain if the galaxies were surrounded by a halo formed by an isothermal gas sphere where the gas pressure and gravity were in *virial equilibrium*. Actually the observed rotation curves suggest that the distribution of dark matter in halos is of the form

$$\rho(R) = \rho_{core} \left(1 + \frac{R^2}{R_{core}^2} \right) \ , \tag{7.36}$$

where the core radius R_{core} and the central density ρ_{core} are two free parameters to be fitted [1].

The observed constant distributions extend out to a distance R_{max}, limited only by the range of observations. If one trusts less accurate methods such as measurements of the relative velocities of galaxy pairs or the kinematics of satellite galaxies, R_{max} could well exceed 100 kpc. The ultimate limit is that R_{max} of each galaxy extends to the confines of its neighbours. In that case one reaches the remarkable conclusion (see Problem 5) that the density of dark and luminous matter equals the critical density of the Universe,

$$\rho = \rho_c \quad , \quad \Omega = 1 \ . \tag{7.37}$$

This result is almost certainly accidental, in particular since it leads to too much mass in galaxy groups and clusters.

Let us now turn to gravitational systems formed by a small number of galaxies. There are examples of such groups in which the galaxies are enveloped in a large cloud of hot gas, visible by its X-ray emission. From the intensity of this radiation the amount of gas can be deduced. Adding this to the luminous matter, the total amount of baryonic matter can be estimated. The temperature of the gas depends on the strength of the gravitational field, from which the total amount of gravitating matter in the system can be deduced.

In the galaxy group HCG62 in the Coma cluster, the Rosat satellite has found a temperature of about 10^7 K [3] which is much higher than what the gravitational field of the visible baryonic matter (galaxies and gas) would produce. One then deduces a baryonic mass fraction of

$$\Omega_B \gtrsim 0.13 \ . \tag{7.38}$$

This cannot be typical of the Universe as a whole if we are to believe in the nucleosynthesis prediction (5.48)

$$0.01 \lesssim \Omega_B \lesssim 0.05 \ . \tag{7.39}$$

On the other hand, this amount of baryonic matter is not enough to close the Universe, or even to agree with the value

$$\Omega_0 = 0.8 \pm 0.3 \tag{7.40}$$

from the QDOT survey [4]. Thus a large dark matter component is missing.

The *virial theorem* for a statistically steady, spherical, self-gravitating cluster of objects, stars or galaxies states that the total kinetic energy of N objects with average random peculiar velocities v equals $-\frac{1}{2}$ times the total gravitational potential energy. If r is the average separation between any two objects of average mass m, the potential energy of each of the possible $N(N-1)/2$ pairings is $-Gm^2/r$. The virial theorem then states that

$$N\frac{mv^2}{2} = \frac{1}{2}\frac{N(N-1)}{2}\frac{Gm^2}{r} \ . \tag{7.41}$$

For a large cluster of galaxies of total mass M and radius R, this reduces to

$$M = \frac{2Rv^2}{G} \; . \qquad (7.42)$$

Thus one can apply the virial theorem to estimate the total dynamic mass of a rich galaxy cluster from measurements of the velocities of the member galaxies and the cluster radius from the volume they occupy. When such analyses have been carried out, taking into account that rich clusters have about as much mass in hot gas as in stars, one finds that baryonic matter accounts for

$$\Omega_B = 0.2 - 0.3 \; . \qquad (7.43)$$

Thus rich galaxy clusters exhibit the same discrepancies as the small HCG62 galaxy cluster: there are too many baryons for the Ω_B-value to represent the average of the Universe, and there are too few to account for the dynamically needed dark matter.

The infall of galaxies in our Local Supercluster (LSC) is manifested by their peculiar velocities toward its centre, which are in the range 150–450 km/s. From this one can derive the local peculiar gravitational field and the mass excess δM concentrated in the Local Supercluster. The mean mass density in the LSC can be obtained by multiplying the mean galaxy mass by the mean number density of galaxies N. Within our radius in the LSC the relative overdensity of bright galaxies is [5]

$$\frac{\delta N}{N} = 2.2 \pm 0.3 \; . \qquad (7.44)$$

If the mass distribution traces the distribution of bright galaxies so that

$$\frac{\delta M}{M} = \frac{\delta N}{N}, \qquad (7.45)$$

one derives a dynamical density parameter value for the LSC in the range [6]

$$0.05 \le \Omega_{LSC} \le 0.45 \; . \qquad (7.46)$$

A useful quantity for comparing different galaxies is the ratio of mass to luminosity,

$$\Upsilon = \frac{M}{L} \; . \qquad (7.47)$$

In solar units, $\Upsilon_\odot$, the value in the solar neighbourhood is about 2.5 with an upper limit of $7\Upsilon_\odot$. Similar values apply to small galaxy groups. However, rich galaxy clusters exhibit much larger Υ ratios, from 300 $\Upsilon_\odot$ for the Coma cluster to 650 $\Upsilon_\odot$. This shows that clusters have their own halo of dark matter which is much larger than the sum of halos of the individual galaxies.

If only a few percent of the total mass of the Universe is accounted for by stars and hydrogen clouds, could baryonic matter in other forms make up dark matter? Gas or dust clouds is the first thing that comes to one's mind. We have already accounted for *hot gas* because it is radiating and therefore visible. Clouds of *cold gas* would be dark but they would not stay cold forever. Unless there exists vastly more cold gas than hot gas, for which there is no reason, this dark matter candidate is insufficient.

It is known that starlight is sometimes obscured by *dust* which in itself is invisible if it is cold and does not radiate. However, dust grains reradiate starlight in the infrared, so they do leave a trace of their existence. But the amount of dust and rocks needed as dark matter would be so vast that it would have affected the composition of stars. For instance, it would have prevented the formation of low-metallicity (population II) stars. Thus dust is not a good candidate.

Snowballs of frozen hydrogenic matter, typical of comets, have also been considered, but they would sublimate with time and become gas clouds. A similar fate exludes *collapsed stars*: they eject gas which would be detectable if their number density would be sufficient for dark matter.

A more serious candidate of baryonic matter are *jupiters* or *brown dwarfs*, stars of mass less than $0.08\,M_\odot$. They also go under the acronym *MACHO* for Massive Compact Halo Objects. They lack sufficient pressure to start hydrogen burning, so their only source of luminous energy is the gravitational energy lost during slow contraction. Such stars would clearly be very difficult to see since they do not radiate. However, if a MACHO passes exactly in front of a distant star, the MACHO would act as a gravitational lens because light from the star bends around the massive object. The intensity of starlight would then be momentarily amplified (on a time scale of weeks or a few months), and this phenomenon of *microlensing* can be detected. The problem is that even if MACHOs were relatively common, one has to monitor millions of stars for one positive piece of evidence. At the time of writing a few microlensing MACHOs have been discovered in the space between Earth and the *Large Magellanic Cloud*[7], but their contribution to Ω_B cannot yet be precisely evaluated. In any case, the nucleosynthesis limit (5.48) applies to this form of baryonic matter.

A more exotic candidate would be *black holes* because they are not luminous and if they are big enough they have a long lifetime, as we saw in Eq. (3.79). They are believed to sit at the centre of galaxies and have masses exceeding $100\,M_\odot$. Whether such a black hole is present in our Galaxy is controversial, and anyway this is not a solution to the galactic rotation curves because dark matter is needed in the haloes, not in the centre of galaxies.

In any case, all the baryonic dark matter candidates conflict with the nucleosynthesis boundary. The shocking conclusion is that the predominant form of matter in the Universe is non-baryonic, and we do not even know what it is composed of ! Thus we are ourselves made of some minor pollutant, a discovery which can well be called the fourth breakdown of the anthropocentric view. I already accounted for the previous three in the first chapter.

The non-baryonic candidates for dark matter can be classified in two groups: *hot dark matter* (HDM) consisting of light particles which were still relativistic at the time of their decoupling, and *cold dark matter* (CDM) particles which are either quite heavy and which therefore decoupled early, or superlight particles with superweak interactions which were never in thermal equilibrium. The most important constraints on dark matter models come from theories of galaxy formation. We shall leave that discussion to Chapter 9.

Problems

1. Derive a value of weak hypercharge $Y = B - L$ for the X boson from reactions (7.7–7.8).
2. Let ρ be the mass of K giants in the central plane $z = 0$ of the Galaxy, and let v_z be the mean random velocity in the z-direction of such stars. Then the pressure associated with their random z-motion at $z = 0$ must be given by

$$p = \rho v_z^2 \ . \tag{7.48}$$

In a steady state, as many stars must cross each level in z during the upward portion of their orbital oscillation as cross in the downward portion, and the partial pressure p must equal the partial weight per unit area of K giants on either side of the plane $z = 0$. (This is the condition of hydrostatic equilibrium of an ideal star gas.) Let D be the effective thickness of the distribution of K giants, and $\langle g_z \rangle$ be the mean value of the z-component of the galactic gravitational field they sample. Then the weight of a column of K giants with unit cross-sectional area is given by $\rho \langle g_z \rangle D$. Apply the condition of hydrostatic equilibrium and show that

$$\langle g_z \rangle = v_z^2 / D \ . \tag{7.49}$$

Compute $\langle g_z \rangle$ from $v_z = 18$ km/s and $D = 300$ pc. [8]
3. The mean free path ℓ of photons in homogeneous interstellar dust can be found from Eq. (1.12) assuming that the radius of dust grains is 10^{-5} cm. Extinction observations indicate that $\ell \approx 1$ kpc at the position of the solar system in the Galaxy. What is the number density of dust grains ? [8]
4. Derive Eq. (7.42). On average the square of the mean random velocity v^2 of galaxies in spherical clusters is three times larger than V^2, where V is the mean line-of-sight velocity displacement of a galaxy with respect to the cluster centre. Calculate M for $V = 1000$ km/s and $R = 1$ Mpc in units of $M_\odot$ [8].
5. Suppose that galaxies have a flat rotation curve out to R_{max}. The total mass inside R_{max} is given by Eq. (7.31) where v may be taken to be 220 km/s. If the galaxy number density is $n = 0.01 \ h^3/\text{Mpc}^3$, show that $\Omega = 1$ when R_{max} is extended out to an average intergalactic distance of $2.5 \ h^{-1}$ Mpc [9].

References

1. S. Tremaine, *Physics Today*, Feb. 1992, p.28.
2. E. Battaner, J. L. Garrido, M. Membrado and E. Florido, *Nature*, **360** (1992) 652.
3. T. J. Ponman and D. Bertram, *Nature*, **363** (1993) 51.
4. D. Lynden-Bell, S. M. Faber, D. Burstein *et al.*, *The Astrophysical Journal*, **326** (1988) 19.
5. G. A. Tammann and A. Sandage, *The Astrophysical Journal*, **294** (1985) 81.
6. P. J. E. Peebles, *The Astrophysical Journal*, **205** (1976) 318.
7. C. Alcock, C-W. Akerlof, R. A. Allsman *et al.*, *Nature*, **365** (1993) 621; E. Auborg, P. Bareyre, S. Bréhin *et al.*, *Nature*, **365** (1993) 623.
8. F. H. Shu, *The Physical Universe*, University Science Books, Mill Valley, CA, 1982.
9. P. J. E. Peebles, *Principles of Physical Cosmology*. Princeton University Press, Princeton, New Jersey, 1993.

8 Cosmic Inflation

In the standard hot Big Bang scenario the early universe is treated as an adiabatically expanding heat bath, and the driving force is the radiation energy density ρ and the positive pressure p. The laws governing its dynamics are Friedman's equations as well as the law of entropy conservation,

$$\dot{s} = 0 . \tag{8.1}$$

In Section 8.1 we shall study a scenario with negative pressure and find that it solves the horizon problem. In 1981 Alan Guth proposed [1] that the flatness problem and the monopole problem could also be made to disappear if the Universe traversed an epoch with negative pressure and terminated it with a huge entropy increase, in violation of the adiabaticity condition (8.1). In Section 8.2 we shall see how this could be possible.

It turned out, however, that the qualitatively nice features of this classical model, now called *old inflation*, could not work quantitatively. Many new models were therefore invented as remedies, in particular by A. Linde, A. Albrecht and P. Steinhardt. It is not clear today which of the current models is correct, if any. In Section 8.3 we shall briefly meet *new inflation* and Linde's *chaotic* version of it [2].

8.1 The Horizon Problem

Recall the definition of the particle horizon in a flat metric in Eq. (2.46),

$$\sigma_p = c \int_0^{t_0} \frac{dt}{S(t)} = c \int_0^{S_0} \frac{dS}{S\dot{S}} . \tag{8.2}$$

In expanding Friedman models the particle horizon is finite, which is at the root of the horizon problem. To see this, let us go back to the derivation of the time dependence of the scale factor $S(t)$ in Eqs. (3.43–46).

At very early times the mass density term in the Friedman equation dominates over the curvature term,

$$\frac{kc^2}{S^2} \ll \frac{8\pi}{3} G\rho \ . \tag{8.3}$$

This permitted us to drop the curvature term and solve for the Hubble parameter,

$$\frac{\dot{S}}{S} = H(t) = \left(\frac{8\pi}{3} G\rho\right)^{\frac{1}{2}} \ . \tag{8.4}$$

Making use of this relation by substituting it in the expression (8.2) for the particle horizon we obtain

$$\sigma_p = c \int_0^{S_0} \frac{\mathrm{d}S}{S^2(\dot{S}/S)}) = \left(\frac{3c^2}{8\pi G}\right)^{\frac{1}{2}} \int_0^{S_0} \frac{\mathrm{d}S}{S^2 \sqrt{\rho}} \ . \tag{8.5}$$

In a radiation-dominated universe ρ scales like S^{-4}, so the integral on the right converges at the lower limit, and the result is that the particle horizon is finite,

$$\sigma_p \propto \int_0^{S_0} \frac{\mathrm{d}S}{S^2 S^{-2}} = S_0 \ . \tag{8.6}$$

Similarly, in a matter-dominated universe ρ scales like S^{-3}, so the integral also converges, now to yield $\sqrt{S_0}$.

Suppose however, that the curvature term or a cosmological constant dominates the Friedman equation at some epoch. Then the condition (3.44) is not fulfilled, on the contrary we have

$$p < -\tfrac{1}{3}\rho c^2 \ . \tag{8.7}$$

Thus the net pressure is negative as in the case of vacuum energy domination in Eq. (3.41). Substituting this into the law of energy conservation (3.33) we find

$$\dot{\rho} < 2\frac{\dot{S}}{S}\rho \ . \tag{8.8}$$

This can easily be integrated to give the S-dependence of ρ,

$$\rho < S^{-2} \ . \tag{8.9}$$

Let us insert this dependence into the integral on the right of Eq. (8.5). We then find

$$\sigma_p \propto \int_0^{S_0} \frac{\mathrm{d}S}{S^2 \sqrt{S^{-2}}} = \int_0^{S_0} \frac{\mathrm{d}S}{S} \ , \tag{8.10}$$

an integral which does not converge at the limit $S = 0$. Thus the particle horizon is not finite in this case.

The lesson of this is that we can get rid of the horizon problem by choosing physical conditions where the net pressure is negative and obeying the condition (8.7), either by having a large curvature term or a dominating cosmological term or some large scalar field which acts as an effective cosmological term. Let us turn to the latter case in the next section.

8.2 Guth's Classical Model

Suppose that the primeval universe was pervaded by a scalar classical field φ at Planck time

$$t_P = \frac{\hbar}{M_P c^2} = 5.31 \times 10^{-44} s .\qquad(8.11)$$

Recall that the Planck mass was defined in Eq. (3.13) as the scale at which quantum effects and gravitation become of equal importance. Since we do not have any theory of quantum gravity, our theoretical description of physical phenomena then breaks down. Thus the earliest time we can meaningfully consider is Planck time.

If the mass m_φ of the scalar quanta carrying the field φ were much lighter than the Planck mass,

$$0 < m_\varphi \ll m_P ,\qquad(8.12)$$

these fields could be considered massless. In fact, the particle symmetry at Planck time is characterized by all other fields than scalar fields being exactly massless. Only when this symmetry is spontaneously broken in the transition to a lower temperature phase, some particles become massive. We have met this situation before in Sections 6.5 and 6.6 where scalar Higgs fields played an important rôle.

Let us introduce the potential $V(\varphi,T)$ of the scalar field at temperature T. Its φ-dependence is arbitrary, but we could take it to be a power function of φ just as we chose to do in Eqs. (6.67) and (6.76). Suppose that the potential at time t_P has a minimum at a particular value φ_P. The Universe would then settle in that minimum given enough time, and the value φ_P would gradually pervade all of spacetime. It would be difficult to observe such a constant field because it would have the same value to all observers, regardless of their frame of motion. Thus the value of the potential $V(\varphi_P,T_P)$ may be considered a property of the vacuum.

Suppose that the minimum of the potential is at $\varphi_P = 0$ in some region of spacetime, and it is non-vanishing,

$$|V(0,T_P)| > 0 .\qquad(8.13)$$

An observer moving along a trajectory in spacetime would notice that the field fluctuates around its *vacuum expectation value*

$$\langle\varphi_P\rangle = 0 ,$$

and the potential energy consequently fluctuates around the mean *vacuum energy* value

$$\langle V(0,T_P)\rangle > 0 .$$

This vacuum energy contributes to the energy density term in Friedman's equation (4.20) which can be written

$$\frac{\dot{S}^2}{S^2} + \frac{kc^2}{S^2} = \frac{8\pi G}{3}[\rho + \langle V(0,T)\rangle] \ . \tag{8.14}$$

The vacuum energy term acts as a repulsive force, just as the cosmological constant term in Eq. (4.28) which Einstein invented to balance the attractive gravitational force. Thus we can make the identification

$$\frac{8\pi G}{3}\langle V_0\rangle \equiv \frac{\lambda}{3} \ , \tag{8.15}$$

where $V_0 = V(0,0)$ is a temperature-independent constant.

As long as the energy density term in Eq. (8.14) is large it drives the expansion, but since $\rho(S)$ decreases there comes a moment when the constant vacuum energy starts to dominate and drive the expansion. The Friedman universe then changes into a de Sitter universe in which $T \simeq 0$. From the solution (4.69) for a flat universe we know that the scale factor then increases exponentially

$$S(t) = e^{Ht} \ . \tag{8.16}$$

Guth named this an *inflationary universe*. From Eq. (8.4) the time scale for inflation is

$$H = \sqrt{\frac{8\pi G}{3}\langle V_0\rangle} \ \propto \ \frac{\sqrt{\hbar c}}{M_P} \simeq (10^{-34}s)^{-1} \ . \tag{8.17}$$

Clearly the cosmic inflation cannot go on forever if we want to arrive at our present slowly expanding Einstein–de Sitter universe. Thus there must be a mechanism to halt the exponential expansion, a *graceful exit*. The freedom we have to arrange this is in the choice of the potential function $V(\varphi,T)$ at different temperatures T.

Suppose that there is a symmetry breaking phase transition from a hot G_{GUT}-symmetric phase dominated by the scalar field φ to a cooler G_s-symmetric phase. As the Universe cools through the critical temperature T_{GUT}, bubbles of the cool phase start to appear and begin to grow. If the rate of bubble nucleation is initially small the Universe supercools in the hot phase, very much like a supercooled liquid which has a state of lowest potential energy as a solid.

We assumed above that the potential $V(\varphi,T_{hot})$ in the hot G_{GUT}-symmetric phase was symmetric around the point $\varphi = 0$ as in the top curve of Fig. 39. Suppose now that $V(\varphi,T)$ develops a new asymmetric minimum as the temperature decreases. At T_{GUT} this second minimum may be located at φ_{GUT}, and the potential may have become equally deep in both minima, as in the middle curve in Fig. 39. Now the Universe could in principle *tunnel* out to the second minimum, but the potential barrier between the minima makes this process slow. Tunnelling through potential barriers is classically forbidden, but possible in quantum physics because quantum laws are probabilistic. If the potential barrier is high or wide, tunnelling is less probable. This is the reason why the initial bubble nucleation can be considered slow.

The lowest curve in Fig. 39 illustrates the final situation when the true minimum has stabilized at φ_0 and the potential energy of this true vacuum is lower than in the original *false vacuum*,

$$V(\varphi_0, T_{cool}) < V(0, T_{hot}) \ .$$

When the phase transition from the supercooled hot phase to the cool phase finally occurs at T_{cool} the latent heat stored as vacuum energy is liberated in the form of radiation and kinetic energy of ultrarelativistic massive scalar particles with positive pressure. At the same time other GUT fields present massify in the process of spontaneous symmetry breaking, filling the Universe suddenly with particles of temperature T_R. The liberated energy is of the order of

$$\langle V_0 \rangle \simeq (kT_R)^4 \ . \tag{8.18}$$

This heats the Universe enormously, from an ambient temperature

$$T_{cool} \ll T_{GUT}$$

to T_R, which is at the T_{GUT} scale. The remaining energy in the φ field is dumped in the entropy which is proportional to T^3. Thus the entropy per particle is suddenly increased by the very large factor

$$Z^3 = \left(\frac{T_R}{T_{cool}} \right)^3 \ . \tag{8.19}$$

This is a non-adiabatic process, grossly violating the condition (8.1).

At the end of inflation the Universe is a hot bubble of particles and radiation in thermal equilibrium. The energy density term in Friedman's equations has become dominant, and the Universe henceforth follows an Einstein–de Sitter type evolution.

The flatness problem is now solved if the part of the Universe which became our Universe was originally homogeneous. We noted in Eq. (7.29) that the present value of the dimensionless 'constant' d has the astoundingly small value 10^{-58}. But in the inflationary scenario its original value was Z^2 times larger. This could well be of the order of unity if the value of Z were

$$Z = \frac{T_R}{T_{cool}} > 10^{29} \ . \tag{8.20}$$

Thus our Universe appears nearly flat because an originally homogeneous region has expanded by a factor

$$e^{H\tau} \simeq 10^{29} \ , \tag{8.21}$$

or $H\tau \simeq 65$. Superimposed on the homogeneity of the pre-inflationary universe there were small fluctuations in the field φ or in the vacuum energy. At the end of inflation these give rise to density fluctuations which are the seeds of later mass structures.

It follows from Eq. (8.21) that the duration of the inflation was

$$\tau \simeq 65 \cdot 10^{-34} s \ . \tag{8.22}$$

Thus the horizon problem is solved, since the initial particle horizon has been blown up by a factor 10^{29} to a size vastly larger than our present Universe. (Note that the realistic particle horizon is not infinite as one would obtain from Eq. (8.10), because the lower limit of the integral is small but non-zero.) In consequence, all the large-scale structures seen today have their common origin in a microscopic part of the Universe long before the last scattering of radiation. The development of Guth's scenario through the pre-inflationary, inflationary and post-inflationary eras is similar to Linde's scenario shown in Fig. 43 except that the vertical scale here grows 'only' to 10^{29}.

When our bubble of spacetime nucleated, it was separated from the surrounding supercooled hot phase by domain walls. When the phase transition finally occurred the enormous amounts of latent heat was released to these walls. The mechanism whereby this heat was transferred to particles in the bubbles was by the collision of domain walls and the coalescence of bubbles. In some models knots or topological defects then remained in the form of monopoles, of the order of one per bubble. Thus the inflationary model also solves the monopole problem by blowing up the size of the region required by one monopole. There remains no inconsistency then with the present observed lack of monopoles.

Although Guth's classical model of cosmic inflation may seem to solve all problems of the hot Big Bang scenario in principle, it fails for quantitative reasons. It turns out to be difficult to tune the rate of nucleation so that the phase transition has time to occur throughout the hot phase. The bubbles of true vacuum grow too slowly, which allows time for the initial density fluctuations to grow too large. Ultimately this causes galaxy formation and clustering to be much more inhomogeneous than what is observed. The remaining false vacuum regions expand too rapidly, increasing their physical volume forever, so that inflation never stops. Thus there is no graceful exit from the inflationary scenario.

The first amendments to Guth's model tried to modify the φ^4-potential of Eqs. (6.67) and (6.76) in such a way that the roll from the false minimum to the true minimum would start very slowly, and that the barrier would be very small or absent. However, this required unlikely finetuning and produced too large density fluctuations. Other attempts to save old inflation may require modifications to Einstein's gravity, which is a more drastic change, but perhaps necessary.

8.3 New Inflation and Linde's Chaotic Model

Guth's model made the rather specific assumption that the Universe started out with the vacuum energy in the false minimum $\varphi = 0$ at time t_P. However, Andrei Linde pointed out that this value as well as any other fixed starting value is as improbable as complete homogeneity and isotropy because of the quantum fluctuations at Planck time (see, for example, reference [2]). Instead the scalar field may have had some random starting value φ_a, which could be assumed to be fairly uniform across the horizon of size M_P^{-1}, changing only by an amount

$$\Delta\varphi_a \simeq M_p \ll \varphi_a \; . \tag{8.23}$$

With time the value of the field would change slowly until it finally reached φ_0 at the true minimum $V(\varphi_0)$ of the potential. This unique bubble became our Universe. This model has been called *new inflation* by its inventors Linde, Albrecht and Steinhardt.

But causally connected spaces are only of size M_P^{-1}, so even the metric of spacetime may be fluctuating from open to closed in adjacent spaces of this size. Thus the Universe can be thought of as a chaotic foam of causally disconnected bubbles in which the initial conditions are different, and which would subsequently evolve into different kinds of universes. Only one bubble would become our Universe, and we could never get any information about the other ones. Andrei Linde called this *chaotic inflation*.

According to Heisenberg's uncertainty relation, at a time scale $\Delta t = \hbar/M_P c^2$ the energy is uncertain by an amount

$$\Delta E > \frac{\hbar}{\Delta t} = M_P c^2 \; . \tag{8.24}$$

Let us for convenience work in units common to particle physics where $\hbar = c = 1$. Then the energy density is uncertain by the amount

$$\Delta\rho = \frac{\Delta E}{(\Delta r)^3} = \frac{\Delta E}{(\Delta t)^3} = M_P^4 \; . \tag{8.25}$$

Thus there is no reason to assume that the potential $V(\varphi_a)$ would be much smaller than M_P^4. We may choose a general parametrization for the potential,

$$V(\varphi) \approx \frac{\kappa\varphi^n}{nM_P^{n-4}} \approx M_P^4 \; , \tag{8.26}$$

where $n > 0$ and $0 < \kappa \ll 1$. This assumption ensures that V does not rise too steeply with φ. For $n = 4$ it then follows that

$$\varphi_a \simeq \left(\frac{4}{\kappa}\right)^{1/4} M_P \gg M_P \tag{8.27}$$

when the free parameter κ is chosen to be very small.

In the simplest field theory coupling a scalar field φ to gravitation, the total energy is of the form

$$\frac{1}{2}\dot{\varphi}^2 + \frac{1}{2}(\nabla\varphi)^2 + V(\varphi) \; . \tag{8.28}$$

If the field φ_a is sufficiently homogeneous and stationary so that

$$\dot{\varphi}_a^2 \ll V(\varphi_a) \; , \quad (\nabla\varphi_a)^2 \ll V(\varphi_a) \; , \tag{8.29}$$

only the $V(\varphi)$ term in Eq. (8.28) is needed. The dynamics can then be described by two equations: Friedman's equations with $V(\varphi)$ replacing the energy density ρ,

$$H^2 + \frac{k}{S^2} = \frac{8\pi}{3M_P^2} V(\varphi) , \tag{8.30}$$

and a differential equation obeyed by scalar fields

$$3H\dot\varphi = -\frac{dV(\varphi)}{d\varphi} . \tag{8.31}$$

The latter originates from the *Klein–Gordon* equation which is the simplest covariant field equation for a scalar field. In the above approximation it is essentially an equation for the 'friction' impeding changes in φ. As S grows the curvature term kS^{-2} in Eq. (8.30) can be neglected just as in the classical model. The ensuing equation then describes an exponentially expanding de Sitter universe.

Initially all spacetime regions of size $H^{-1} = M_P^{-1}$ would contain inhomogenities inside their respective event horizons. At every instant during the inflationary de Sitter stage an observer would see himself surrounded by a black hole with event horizon H^{-1} (but remember that 'black hole' really refers to a static metric). Within a time of the order of H^{-1} all inhomogenities would have traversed the Hubble radius. Thus they would not affect the physics inside the de Sitter universe which would be getting increasingly homogeneous and flat. On the other hand, the Hubble radius is also receding exponentially, so if we want to achieve homogeneity it must not run away faster than the inhomogenities.

Combining Eqs. (8.26), (8.30), and (8.31), we obtain an equation for the time dependence of the scalar field,

$$\frac{1}{2}\dot\varphi^2 = \frac{n^2 M_P^2}{48\pi\varphi^2} V(\varphi) . \tag{8.32}$$

Let us study the solution of this equation in the case where $n = 2$ and the potential is given by Eq. (6.64) and Fig. 36,

$$V(\varphi) = \frac{1}{2}m_\varphi^2 \varphi^2 . \tag{8.33}$$

The time dependence of the field is then

$$\varphi(t) = \varphi_a - \frac{m_\varphi M_P}{2\sqrt{3\pi}}t \equiv \varphi_a(1 - t/\tau) , \tag{8.34}$$

where τ is the characteristic time scale of the expansion. At early times when $t \ll \tau$ the scalar field remains almost constant, changing only slowly from φ_a to its ultimate value φ_0. The scale factor then grows quasiexponentially as

$$S(t) = S(t_a)\exp(Ht - \frac{1}{6}m_\varphi^2 t^2) , \tag{8.35}$$

with H given by

$$H = 2\sqrt{\frac{\pi}{3}}\frac{m_\varphi}{M_P}\varphi_a . \tag{8.36}$$

As the field approaches φ_0 the Universe slowly rolls down into the true vacuum at $V(\varphi_0)$, and the inflation ends in graceful exit.

At time τ the Universe has expanded from a linear size $S(t_a)$ to

$$S(\tau) \simeq S(t_a) \exp(H\tau) = S(t_a) \exp\left(\frac{4\pi\varphi_a^2}{M_P^2}\right) \ . \tag{8.37}$$

For instance, a universe of linear size equal to the Planck length $S(t_a) \simeq 10^{-33}$ cm has grown to

$$S(\tau) \simeq S(t_a) \exp\left(\frac{4\pi M_P^2}{m_\varphi^2}\right) \ . \tag{8.38}$$

For a numerical estimate we need a value for the mass m_φ of the scalar particle. This is not known, but we can make use of the condition that the chaotic model must be able to form galaxies of the observed sizes. The anisotropy of the CMB is, as we have seen, of the order of $\delta T/T \approx 10^{-5}$, and the Sachs–Wolfe effect relates this to mass fluctuations of the order of $\delta \approx 10^{-4}$ averaged within one Hubble radius. Then the scalar mass must be of the order of magnitude

$$m_\varphi \simeq 10^{-4} M_P \ . \tag{8.39}$$

Inserting this estimate into Eq. (8.38) we obtain the completely unfathomable scale

$$S(\tau) \simeq 10^{-33}\text{cm} \ \exp(4\pi \cdot 10^8) \simeq 10^{5.5 \times 10^8}\text{cm} \ . \tag{8.40}$$

It is clear that all the problems of the standard Big Bang model discussed in Section 7.2 then disappear. The homogeneity, flatness and isotropy of the Universe turn out to be consequences of the scalar field φ having been large enough in a region of size M_P^{-1} at time t_P. The inflation started in different causally connected regions of spacetime 'simultaneously' to within 10^{-43}s, and it ended at about 10^{-35}s. Our part of that region was extremely small. Since the curvature term in Friedman's equations decreased exponentially, the end result is exactly as if k had been zero to start with. A picture of this scenario is shown in Fig. 43.

Let us now address the question of whether the initial conditions could reasonably be expected to occur so that inflation would start. We noted that at Planck time the field φ was indefinite by M_p, at least. Over a sufficiently long time all fluctuations average to zero so that the classical vacuum appears empty and devoid of properties.

The fate of a bubble of spacetime clearly depends on the value of φ_a. Only when it is large enough will inflationary expansion commence. If φ_a is very much larger than M_P, Eq. (8.36) shows that the rate of expansion is faster than the time scale τ,

$$H \gg 2\sqrt{\frac{\pi}{3}} m_\varphi \simeq \frac{2}{\tau} \ . \tag{8.41}$$

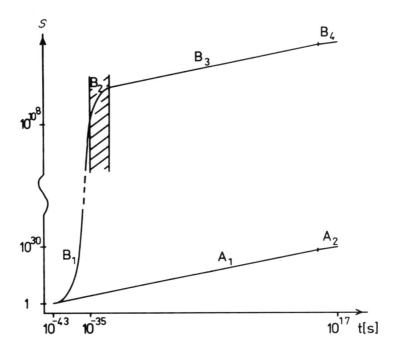

Fig. 43 Evolution of the scale S of the Universe since Planck time in (A) Friedmann models and (B) inflationary expansion. During the epoch B_1 the Universe expands exponentially, and during B_2 the inflation ends by reheating the Universe. After the graceful exit from inflation the Universe is radiation-dominated along B_3, just as in A_1, following a Friedmann expansion. The sections B_4 and A_2 are matter-dominated epochs

Although the wavelengths of all fields then grow exponentially, the change $\Delta\varphi$ in the value of the scalar field itself may be small. In fact, when the wavelengths have reached the size of the Hubble radius H^{-1}, all changes in φ are impeded by the friction $3H\dot{\varphi}$ in Eq. (8.31), and fluctuations of size $\delta\varphi$ freeze to an average non-vanishing amplitude of

$$|\delta\varphi(x)| \simeq \frac{H}{2\pi} .\tag{8.42}$$

In consequence the vacuum no longer appears empty and devoid of properties. The quantum fluctuations remaining in the scalar field will cause that the energy is dumped into entropy at slightly fluctuating times. Thus the Universe will contain entropy fluctuations as seeds of later density fluctuations.

Since the part of the pre-inflationary universe which was inflated to become ours was so small, it may be considered as just one bubble in a foam of bubbles having different fates. In Linde's chaotic model each original bubble has grown in one e-folding time $\tau = H^{-1}$ to a size comprising e^3 mini-universes, each of diameter H^{-1}. In half of these mini-universes, on the average, the value of φ may be large enough for inflation to continue, and in one half it may be too small. In the next e-folding time the same pattern is repeated. Linde has shown that in those parts of spacetime where φ grows continuosly the volume of space grows by a factor

$$e^{(3 - \ell n\ 2)Ht}\ , \tag{8.43}$$

whereas in the parts of spacetime where φ does not decrease the volume grows by the factor

$$\frac{1}{2}e^{3Ht}\ . \tag{8.44}$$

Since the Hubble parameter is proportional to φ, most of the physical volume must come from bubbles in which φ is maximal,

$$\varphi \simeq M_P^2/m_\varphi\ . \tag{8.45}$$

But there must also be an exponential number of bubbles in which φ is smaller. Those bubbles are the possible progenitors of universes of our kind. In them, φ attains finally the value corresponding to the true minimum $V(\varphi_0)$, and an Einstein–de Sitter type evolution takes over. Elsewhere the inflatoric growth continues forever. Thus we live in a Universe which is a minuscule part of a steady-state eternally inflating Meta-Universe which has no end, and therefore it also has no beginning. (If you accept this view, there is no point in reading Chapter 10 !) There is simply no need to turn inflation on in the first place.

During inflation each bubble is generating new spacetime to expand into, as required by general relativity, rather than expanding into pre-existing spacetime. In these de Sitter spacetimes the bubble wall appears to an observer as a surrounding black hole. Two such expanding bubbles are causally disconnected, so they can neither collide nor coalesce. Thus the mechanism of vacuum energy release and transfer of heat to the particles created in the phase transition is not by bubble collisions as in the classical model. Instead, the rapid oscillations of the classical field φ decay away by particle production as the Universe settles in the true minimum. The potential energy then thermalizes at some temperature of the order of T_{GUT}.

Unfortunately the mechanism for baryosynthesis described in Section 7.1 then no longer works. The reason is that the baryon–antibaryon asymmetry produced at the GUT phase transition is subsequently washed out when the Universe reheats at the end of inflation. Thus baryosynthesis is still in the need of an explanation. This could very well be a CP-violating mechanism of the kind discussed, however not associated with the GUT phase transition, but with a later one at a lower temperature.

The inflationary models make two important predictions. Firstly, the Universe should today be critical,

$$\Omega_0 = 1\ . \tag{8.46}$$

Direct evidence on the largest observable scales indicate a value for the deceleration parameter q_0 of ≈ 0.5 [3], implying that Ω_0 could be one. Thus not only do we observe that there is too little luminous matter to explain the dynamical behaviour of matter, we also have a precise theoretical specification for how much matter there should be. This links the dark matter problem to the hypothesis of inflation.

The second prediction is that the initial spectrum of primordial density fluctuations should be of the scale-invariant Harrison–Zel'dovich form. We shall come back to this condition in the next chapter.

Problems

1. Derive Eqs. (8.34) and (8.35).
2. Derive $\varphi(t)$ for a potential $V(\varphi) = \frac{1}{4}\lambda\varphi^4$.
3. Suppose that the scalar field averaged over the Hubble radius H^{-1} fluctuates by the amount ψ. The field gradient in this fluctuation is $\nabla\psi = H\psi$ and the gradient energy density is $H^2\psi^2$. What is the energy of this fluctuation integrated over the Hubble volume? Use the timescale H^{-1} for the fluctuation to change across the volume and the uncertainty principle to derive the minimum value of the energy. This is the amount by which the fluctuation has stretched in one expansion time. [4]
4. Material observed now at redshift $z = 1$ is at present distance H_0^{-1}. The recession velocity of an object at coordinate distance x is $\dot{S}x$. Show that the recession velocity at the end of inflation is

$$\dot{S}x = \frac{H_0 S_0 x z_r}{\sqrt{z_{eq}}} \; , \tag{8.47}$$

where z_r is the redshift at the end of inflation. Compute this velocity. The density contrast has grown by the factor z_r^2/z_{eq}. What value did it have at the end of inflation since it is now $\delta \approx 10^{-4}$ at the Hubble radius?[4]

References

1. A. H. Guth, *Physical Review D*, **23** (1981) 347.
2. A. Linde, *Particle Physics and Inflationary Cosmology*, Harwood Academic Publishers, London, 1990.
3. K. I. Kellermann, *Nature*, **361** (1993) 134.
4. P. J. E. Peebles, *Principles of Physical Cosmology*. Princeton University Press, Princeton, New Jersey, 1993.

9 Galaxy Formation

Although we have reached an impressively accurate description of the dynamics of the Universe assuming homogeneity and adiabaticity, we still have to explain where and how matter began to cluster, ultimately to form the perhaps 10^9 galaxies (within the Hubble radius) and the large-scale structures. The baryonic matter did not start to form atoms until recombination, but the clustering of baryons could have started from the moment the Universe became matter-dominated. The time and size scales are important constraints to galaxy formation models, as are the curious patterns of filaments, sheets and voids on the very large scales we have seen in Fig. 1 (Plate 2, frontispiece). Very important constraints furnish the observations of CMB anisotropy and of the galaxy peculiar velocity field.

Nevertheless, there exists no convincing theory of galaxy formation. We must therefore content ourselves with a basic discussion of possible candidate models and describe their problems. Section 9.1 familiarizes us with the formalism of density fluctuations in the hot plasma of electrons, photons and protons. In Section 9.2 we turn to models of cold dark matter which depend on the existence of hypothetical particles. In Section 9.3 we discuss neutrinos as hot dark matter. The properties of dark matter discussed here are those required for galaxy formation, which is not quite the same as the properties required for dynamical dark mass in galactic haloes which were discussed in Section 7.3.

9.1 Density Fluctuations

The distribution of matter in the Universe can, to some approximation, be described by the hydrodynamics of a viscous, non-static fluid. In such a medium there naturally appear random fluctuations around the mean density $\bar{\rho}(t)$, manifested by compressions in some regions and rarefactions in other regions. An ordinary fluid is dominated by the material pressure, but in the fluid of our Universe radiation pressure and gravitational attraction are competing. This makes the physics different from ordinary hydrodynamics: regions of overdensity are gravitationally amplified and may grow into large inhomogenities, depleting adjacent regions of underdensity.

Let us denote the local density $\rho(r,t)$ at comoving spatial coordinate $\mathbf{r}$ and world time t.

Then the fractional departure at **r** from the spatial mean density $\bar{\rho}(t)$, or the *mass density contrast*, is

$$\delta(\mathbf{r},t) = \frac{\rho(\mathbf{r},t) - \bar{\rho}(t)}{\bar{\rho}(t)} \ . \tag{9.1}$$

An arbitrary pattern of fluctuations can be mathematically described by an infinite sum of independent waves, each with its characteristic wavelength λ or wave number k and its amplitude δ_k. The sum can be formally expressed as a *Fourier expansion* for the density contrast

$$\delta(\mathbf{r},t) \propto \sum \delta_k(t) e^{i\mathbf{k}\cdot\mathbf{x}} \ , \tag{9.2}$$

where **k** is the wave vector. A density fluctuation can also be expressed in terms of the mass M moved within one wavelength, or rather within a sphere of radius λ, thus $M \propto \lambda^3$. It follows that the wave number k depends on mass as

$$k = \frac{2\pi}{\lambda} \propto M^{-1/3} \ . \tag{9.3}$$

A useful quantity is the dimensionless *mass autocorrelation function*

$$\xi(r) = \langle \delta(\mathbf{r}_1)\delta(\mathbf{r}+\mathbf{r}_1) \rangle \propto \sum \langle |\delta_k(t)|^2 \rangle e^{i\mathbf{k}\cdot\mathbf{r}} \ . \tag{9.4}$$

which measures the correlation between the density contrasts at two points **r** and $\mathbf{r}_1$. The *powers* $|\delta_k|^2$ define the *power spectrum* of the rms mass fluctuations,

$$P(k) = \langle |\delta_k(t)|^2 \rangle \ . \tag{9.5}$$

Thus the autocorrelation function $\xi(r)$ is the Fourier transform of the power spectrum. We have already met a similar situation in the context of CMB anisotropies where the waves represented temperature fluctuations on the surface of the surrounding sky, and the powers a_ℓ^2 were coefficients in the Legendre polynomial expansion Eq. (5.19).

With the lack of more accurate knowledge of the power spectrum one assumes for simplicity that it is specified by a power law

$$P(k) \propto k^n \ , \tag{9.6}$$

where n is the *spectral index*. Combining this with Eq. (9.3) one sees that each mode δ_k is proportional to some power of the characteristic mass M. The inflationary models predict that the power spectrum is almost scale-invariant as the fluctuations cross the Hubble radius. This is the Harrison–Zel'dovich spectrum, for which $n = 1$ ($n = 0$ would correspond to white noise).

Some information about the spectral index can be derived from constraints set by black holes. From Eq. (3.16) we see that black holes less massive than 10^{12} kg will have evaporated already within 10 Gyr. If the fluctuations are non-linear on a scale larger than 10^{12} kg there would be too many black holes evaporating now and contributing to the CMB. Only very large black holes which do not evaporate fast may exist. Thus to be consistent with CMB, black holes and galaxy masses, the spectrum must be close to the Harrison–Zel'dovich form. The COBE measurements are consistent with a spectral index in the range 0.5–1.6.

Theoretical models of density fluctuations can be specified by the amplitudes δ_k of the autocorrelation function $\xi(r)$. In particular, if the fluctuations are Gaussian, they are completely specified by the power spectrum $P(k)$. The models can then be compared to the real distribution of galaxies and galaxy clusters. Suppose that the galaxy number density in a volume element dV is n_G, then one can define the probability of finding a galaxy in a random element as

$$dP = n_G dV .\tag{9.7}$$

If the galaxies are distributed independently, for instance with a spatially homogeneous Poisson distribution, the joint probability of having one galaxy in each of two random volume elements dV_1, dV_2 is

$$dP_{12} = n_G^2 dV_1 dV_2 .\tag{9.8}$$

There is then no correlation between the probabilities in the two elements. However, if the galaxies are clustered on a characteristic length r_c the probabilities in different elements are no longer independent but correlated. The joint probability of having two galaxies with a relative separation r can then be written

$$dP_{12} = n_G^2[1 + \xi(r/r_c)]dV_1 dV_2 ,\tag{9.9}$$

where $\xi(r/r_c)$ is the *two-point correlation function* for the galaxy distribution. This can be compared to the autocorrelation function (9.4) of the theoretical model.

Analyses of galaxy clustering show [2] that for distances

$$10 \text{ kpc} \lesssim hr \lesssim 10 \text{ Mpc}\tag{9.10}$$

a good empirical form for the two-point correlation function is

$$\xi(r/r_c) = (r/r_c)^{-\gamma} ,\tag{9.11}$$

with the parameter values $r_c = 5.4h^{-1}$ Mpc, $\gamma = 1.77$.

As we have seen from Friedman's equations, the force of gravity makes a homogeneous matter distribution unstable; it either expands or contracts. Thus if the matter density fluctuates, also the cosmic scale factor will be a fluctuating function $S(r,t)$ of position and time. In particular, while the Universe is expanding in the mean, there may be overdense regions where the gravitational forces dominate over pressure forces, causing matter to contract locally and to attract surrounding matter. In other regions where the pressure forces dominate, the fluctuations move as sound waves in the fluid, damped by its viscosity and transporting energy from one region of space to another.

The dividing line between these two possibilities can be found by a classical argument. Let the time of free fall to the centre of an inhomogeneity in a gravitational field of strength G be

$$t_G = 1/\sqrt{G\bar{\rho}} \; . \tag{9.12}$$

Sound waves propagate with velocity c_s, thus they move one wavelength λ in the time

$$t_S = \lambda/c_S. \tag{9.13}$$

If t_G is shorter than t_S the overdensities will contract, and the amplitude of the fluctuations will grow by attracting surrounding matter. In the opposite case, the fluctuations will move with constant amplitude as sound waves. Setting $t_G = t_S$ we find the limiting wavelength $\lambda = \lambda_J$, called the *Jeans wavelength* after Sir James Jeans (1877–1946),

$$\lambda_J = \sqrt{\frac{\pi}{G\bar{\rho}}} c_S \; . \tag{9.14}$$

Actually the factor $\sqrt{\pi}$ was not present in the above Newtonian derivation, it comes from a more exact hydrodynamic treatment (see for instance reference [2]). The mass contained in a sphere of radius λ_J is called the *Jeans mass*,

$$M_J = \tfrac{4}{3}\pi\lambda_J^3\rho \; . \tag{9.15}$$

Let us distinguish between two types of fluctuations, *adiabatic* or *curvature* fluctuations and *isocurvature* fluctuations. In adiabatic fluctuations the local number densities of baryons and photons fluctuate simultaneously. By the principle of covariance, fluctuations in the energy-momentum tensor imply simultaneous fluctuations in energy density and pressure, and by the equivalence principle, variations in the energy-momentum tensor are equivalent to variations in the curvature. Curvature fluctuations can have been produced early as irregularities in the metric, and then can have been blown up by inflation far beyond the Hubble radius. Thus adiabatic fluctuations are a natural consequence of cosmic inflation. We can measure the irregularities in the metric by the curvature radius r_U defined in Eq. (3.26). If r_U is less than the linear dimensions d of the fluctuating region, it will collapse as a black hole. To establish the relation between the curvature of the metric and the size of the associated mass fluctuation requires the full machinery of general relativity, which is beyond our ambitions.

In isocurvature scenarios the matter density starts out homogeneous, so that the curvature of the metric is unperturbed. Fluctuations in some particle species then cause the matter density to vary independently of the pressure. Since density and pressure are related by the equation of state, this corresponds to spatial variations in the local equation of state. Particles cannot influence events outside the horizon, but in inflationary scenarios the horizon grows so enormously that primordial isocurvature fluctuations will subsequently be inflated far outside $r_H = H^{-1}$. Note that r_H is of the order of the particle horizon in a universe following a Friedman expansion, whereas it is much smaller than the horizon in inflationary scenarios.

In an expanding universe we may further distinguish between fluctuations in the linear regime, $|\delta_k| < 1$, where the size of the fluctuations and the wavelengths grow linearly with the scale S, and the non-linear regime, $|\delta_k| > 1$, where the density fluctuations grow faster, with the power S^3, at least. We can also express the density contrast in terms of the linear size d of the region of overdensity normalized to the curvature radius,

$$\delta \approx \left(\frac{d}{r_U}\right)^2 . \tag{9.16}$$

In the linear regime r_U is large, so the Universe is flat. At the epoch when d is of the order of the Hubble radius, the density contrast is

$$\delta_H \approx \left(\frac{r_H}{r_U}\right)^2 , \tag{9.17}$$

free streaming can leave the region and produce the CMB anisotropies. Structures have formed when $d \ll r_H$, thus when $\delta \ll 1$. Although δ may be very small, the fluctuations may have grown by a very large factor because they started early on (see Problem 4 in Chapter 8).

The present roughly homogeneous state of the Universe cannot have grown out of primeval chaos, because a chaotic universe can grow homogeneous only if the initial conditions are incredibly well fine-tuned. Vice versa, a homogeneous universe will grow more chaotic, because the standard model is gravitationally unstable.

After the graceful exit from inflation the Universe enters the regime of Friedman expansion, during which the Hubble radius overtakes the inflated regions. Thus if there were pre-inflationary fluctuations they will cross the post-inflationary Hubble radius and come back into vision with a wavelength corresponding to the size of the Hubble radius at that moment. This is illustrated in Fig. 44.

The subsequent evolution depends on what amplitude the fluctuations have then. Before recombination the baryonic Jeans mass is some 30 times larger than the mass of baryons within the Hubble radius, so if there exist non-linear modes they are outside it. Well inside the Hubble radius the fluctuations may start to grow as soon as the Universe becomes matter-dominated which occurs at time $t_{eq} = 1400\,(\Omega_0 h^2)^{-3/2}$ yr. Fluctuations which enter in the non-linear regime, where the ratio in Eq. (9.17) is large, collapse rapidly into black holes before pressure forces have time to respond.

Adiabatic fluctuations lead to gravitational collapse if the mass scale is so large that the radiation does not have time to diffuse out of one Jeans wavelength within the time t_{eq}. Large wavelength fluctuations will grow with the expansion as long as they are in the linear regime. Short wavelength mass fluctuations with $\lambda \approx r_U \ll r_H$ can break away from the general expansion and collapse to bound systems of the size of galaxies or clusters of galaxies. Still shorter wavelengths (large k in a comoving frame) oscillate in the linear regime as sound waves, never reaching gravitational collapse. Random waves moving through the medium with the speed of sound v erase all perturbations with wavelengths less than vt_{eq}.

During the radiation-dominated epoch and until recombination the buildup and growth of baryonic density perturbations is inhibited by the tight electromagnetic coupling of electrons, protons and photons which hinders them from travelling very far. This causes the fluid to be viscous, so that density perturbations on small scales are strongly damped by photon diffusion and thermal conductivity as soon as they arise, and on large scales there is no coherent energy transport. If the fluid is composed of some

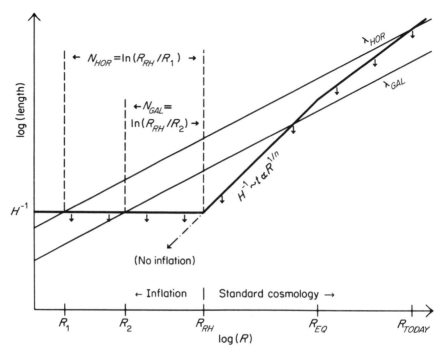

Fig. 44   The evolution of the physical size of the comoving scale or wavelength λ, and of the Hubble radius H^{-1} as functions of the scale of the Universe R (corresponding to S in our notation). In the standard non-inflationary cosmology a given scale crosses the horizon but once, while in the inflationary cosmology all scales begin sub-horizon sized, cross outside the Hubble radius ('good bye') during inflation, and re-enter ('hello again') during the post-inflationary epoch. Note that the largest scales cross outside the Hubble radius first and re-enter last. The growth in the scale factor, $N=\ln{(R_{RH}/R)}$, between the time a scale crosses outside the Hubble radius during inflation and the end of inflation is also indicated. For a galaxy, $N_{GAL}=\ln{(R_{RH}/R_1)}\approx 45$, and for the present horizon scale, $N_{HOR}=\ln{(R_{RH}/R_2)}\approx 53$. Causal microphysics operates only on scales less than H^{-1}, indicated by arrows. During inflation H^{-1} is a constant, and in the post-inflation era it is proportional to $R^{1/n}$, where $n=2$ during radiation-domination, and $n=3/2$ during matter-domination. Reproduced from reference [1] by permission of E. W. Kolb and M. S. Turner

non-baryonic particle species, it need not be viscous to all these particles, depending on their interactions.

The situation changes dramatically at recombination, when all the free electrons suddenly disappear, captured into atomic Bohr orbits, and the radiation pressure almost vanishes. This occurs at time 180 000 $(\Omega_0 h^2)^{-1/2}$ yr after Big Bang. Now the density perturbations which have entered the Hubble radius can grow with full vigour.

At some stage the hydrogen gas in gravitationally contracting clouds heats up enough to become ionized. The free electrons and photons then build up a radiation pressure halting further collapse. The state of such clouds today depends on how much mass and time there was available for their formation. Small clouds may shrink rapidly, radiating their gravitational binding energy and fragmenting. Large clouds shrink slowly and cool by the mechanism of electron Thomson scattering. As the recombination temperature is approached the photon mean free paths become larger, so that radiation can diffuse out of overdense regions. This damps the growth of inhomogeneities.

Only clouds exceeding the Jeans mass stabilize and finally virialize. It is intriguing (but perhaps an accident) that the Jeans mass just after recombination is about $10^5 M_\odot$, the size of globular clusters ! Galaxies have masses of the order of $10^{12} M_\odot$ corresponding to fluctuations of order $\delta \simeq 10^{-4}$ as they cross the horizon. We already made use of this fact to fix the mass m_φ of the scalar field in Eq. (8.39). Gravitational clustering in open universes with $\Omega_0 < 1$ ends at time $\Omega_0 t_0$, where t_0 denotes the present age as usual.

The timetable for galaxy and cluster formation is restricted by two important constraints. At the very earliest, the Universe has to be large enough to have space for the first formed structures. If these were galaxies of the present size, their number density can be used to estimate how early they could have been formed. We leave this for a Problem.

The present density of the structures also sets a limit to the formation time. The density contrast at formation must have exceeded the mean density at that time, and since then the contrast has increased like S^3. Thus rich clusters, for instance, cannot have been formed much earlier than at

$$S = 1 + z \approx 2.5 \Omega^{-1/3} . \tag{9.18}$$

It seems that all the present structure was in place already at $z = 5$. This does not exclude that the largest clusters are still collapsing today.

As a result of mass overdensities, the galaxies influenced by the ensuing fluctuations in the gravitational field will acquire peculiar velocities. One can derive a relation between the mass autocorrelation function and the rms peculiar velocity, see reference [2]. If one takes the density contrast to be $\delta = 0.3$ for rms fluctuations of galaxy number density within a spherical volume radius $30\ h^{-1}$ Mpc, and if one further assumes that all mass fluctuations are uncorrelated at larger separations, then the acceleration caused by the gravity force of the mass fluctuations would predict deviations from a homogeneous peculiar velocity field in rough agreement with observations in our neighbourhood. Much larger density contrast would be in marked disagreement with the standard model and with the velocity field observations.

So far we have considered the fluctuating mass density to be baryonic. However, nucleosynthesis considerations have shown that the amount of baryonic matter is very small, so if the galaxies have arisen from primordial density fluctuations in a purely baryonic medium, the amplitude of the fluctuations must have been very large. But the amplitude of fluctuations in radiation must then also be very large, because of adiabaticity. This leads to intolerably large CMB anisotropies today. Thus galaxy formation in purely baryonic matter is ruled out by this argument alone.

As we have seen in Chapter 7, there must exist dark matter in the galactic haloes in order to explain the galaxy rotation curves. Moreover, the value $\Omega_0 = 1$ expected in inflationary models requires at least an order of magnitude more dark matter than baryonic matter, including MACHOs. Galaxy formation will then also depend on the nature of dark matter and on the interactions of the particles constituting it. Let us therefore continue the study of non-baryonic dark matter models.

9.2 Cold Dark Matter

Particles which were very slow at time t_{eq}, when the radiation density equalled the matter density and galaxy formation started, are candidates for *cold dark matter* (CDM). If these particles are massive and have weak interactions, so called WIMPs, *weakly interacting massive particles*, they became non-relativistic much earlier than the leptons and decoupled from the hot plasma. For instance, the supersymmetric models contain at least three such particles of which one or two (the *photino* and the *Zino*) or a a linear combination of them (the *gaugino*) could serve. Laboratory searches have not (yet?) found them up to a mass of about 20 GeV. Neutrinos may also be CDM candidates if very heavy ones, $m_\nu > 45$ GeV, exist.

Alternatively, the CDM particles may be very light if they have some superweak interactions, in which case they froze out early when their interaction rate became smaller than the expansion rate, or they never even attained thermal equilibrium. Candidates in this category are the *gravitino* which is a supersymmetric partner to the graviton, and the *axion* which is an almost massless boson related to a slightly broken baryon number symmetry. Both of these appear naturally in supersymmetric theories.

All CDM candidates have in common that they are hitherto unobserved particles which only exist in some theories. A signal worth looking for would be monoenergetic photons from their annihilation

$$X_{DM} + \bar{X}_{DM} \to 2\gamma. \tag{9.19}$$

Several experiments are planned or under way to observe these photons if they exist.

The seeds of CDM fluctuations may have been generated in some early phase transition long before recombination. The standard assumption is that the linear mass fluctuations were randomly distributed, Gaussian, so that the power spectrum had the scale-invariant Harrison–Zel'dovich form $P(k) \propto k$ when the fluctuations entered the horizon. From the time t_{eq} until recombination they grew unhindered by the viscosity felt by the baryons in the electron and photon plasma. Note that matter-domination starts much earlier in the presence of dark matter than if there were only baryons.

When the WIMPs decouple from the plasma they can stream freely out of overdense regions and into underdense regions, thereby erasing all small inhomogenities. This defines the characteristic length and mass scales for freely streaming particles of mass m_{dm},

$$\lambda_{fs} \simeq 40 \left(\frac{30 \text{ eV}}{m_{dm}} \right) \text{Mpc}, \tag{9.20}$$

$$M_{fs} \simeq 3 \times 10^{15} \left(\frac{30 \text{ eV}}{m_{dm}} \right)^2 M_\odot. \tag{9.21}$$

After recombination the baryons fall into the CDM potential wells. A few expansion times later, the baryon perturbations catch up with the WIMPs, and both then grow together until $\delta > 1$, when perturbations become Jeans unstable, collapse and virialize. The amplitude of radiation, however, is unaffected by this growth, so the CMB anisotropies remain at the level determined by the baryonic fluctuations just before recombination. This is illustrated in Fig. 45.

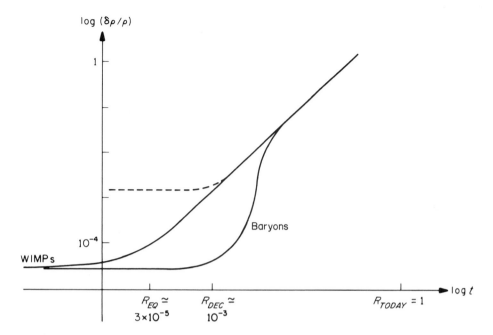

Fig. 45 The evolution of density fluctuations $\delta\rho/\rho$ in the baryon and WIMP components. The perturbations in the WIMPs begin to grow at the epoch of matter-radiation equality. However, the perturbations in the baryons cannot begin to grow until just after decoupling, when baryons fall into the WIMP potential wells. Within a few expansion times the baryon perturbations 'catch up' with the WIMP perturbations. The dashed line shows where the baryonic density fluctuations would have to start if dark matter were purely baryonic. Reproduced from reference [1] by permission of E. W. Kolb and M. S. Turner

The non-baryonic WIMPs should then be found today in the haloes of the galaxies together with the baryonic MACHOs. On the other hand, if the WIMP overdensities only constituted early potential wells for the baryons, but did not cluster so strongly, most WIMPs would have leaked out now into the intergalactic space. In that case the WIMP distribution is more uniform than the galaxy distribution, so that light does not trace mass.

If the non-baryonic dark matter in our galactic halo were constituted by WIMPs at a sufficient density, they should also be found inside the Sun where they lose energy in elastic collisions with the protons, and ultimately get captured. They could then contribute to the energy transport in the Sun, modifying solar dynamics noticeably. So far this possible effect has not been observed.

The WIMPs would of course also traverse terrestrial particle detectors with a typical virial velocity of the order of 200 km/s, and perhaps leave measurable recoil energies in the elastic scattering with protons. The proof that the recoil detected was due to a particle in the galactic halo would be the annual modulation of the signal. Because of the motion of the Earth around the Sun, the signal should have a maximum in June and a minimum in December. Several experiments trying to detect such signals are currently running or being planned, but so far the absence of signals only permits us to set upper limits to the WIMP flux.

The lightest WIMPs are slow enough at time t_{eq} to be bound in perturbations on galactic scales. Although the CDM models do produce galaxies naturally, they underproduce galaxy clusters and supergalaxies of mass scale $10^{15}M_\odot$. CDM therefore predicts a 'down–top' scenario where small-scale structures are produced first and large-scale structures have to be assembled from them later. Unfortunately for the model, there is not time enough to achieve this within the known age of the Universe.

This question can be judged by comparing a simulated mass autocorrelation function with the observed galaxy number two-point correlation function. If the two correlation functions are identical, this implies that light traces mass exactly. If not, they are biased to a degree described by the *bias parameter b*. The physical meaning of the bias is that the amplitude of the fluctuations in dynamical mass would be a factor b^{-1} times the fluctuations in galaxy number density, thus $b = 1$ corresponds to unbiased. The need for the bias parameter comes from the conflict between the assumption $\Omega_0 = 1$ inherent in the CDM model, and the small-scale dynamical estimates of dark matter in Section 7.3. Clearly, galaxies are much more strongly clustered than the dark matter distribution from which they originated, and galaxy clusters are even more strongly clustered.

On distance scales of 20 h^{-1} Mpc ('small distances') computer simulations in adiabatic CDM models match the galaxy two-point correlation function with a bias parameter chosen in the range $1.7 \stackrel{<}{\sim} b \stackrel{<}{\sim} 3.0$. This implies a relatively important amplitude in the baryonic component, leading to a CMB anisotropy of $\Delta T/T \simeq 10^{-5}$ on an angular scale of $1°$ in some disagreement with the South Pole results quoted in Eq. (5.20) [6]. Another indicator of important biasing on this scale is the peculiar velocity field of nearby galaxies relative to the Hubble flow. If dark matter were tracing light, the peculiar velocities would have to be of the order of 1000 km/s whereas the observed velocities are only about 400 km/s.

However, the existence of a well-fitting two-point correlation function on small scales does not preclude the existence of lumpiness on larger scales. Astronomers have recently observed several surprisingly large structures which cannot very well arise in either adiabatic or isocurvature CDM models which have insufficient power at intermediate and large scales. We have already referred to the 'Great Attractor', a hypothetical overdensity of mass about $5.4 \times 10^{16}M_\odot$ in the direction of the *Hydra-Centaurus* cluster, but far behind it, at a distance of some 44 h^{-1} Mpc [3]. The evidence for it is the large scale bulk flows of matter observed within our supergalaxy indicating $b < 1.4$. The COBE measurement at distance scales of several 100 Mpc indicate a bias in the range 0.7–1.2 when the spectral index is assumed to be $n = 1$. [7].

The distribution of galaxies in two-dimensional pictures shows that the galaxies form long filaments separating large underdense voids with diameters up to 60 h^{-1} Mpc. As three-dimensional pictures of the Universe such as Fig. 1 (Plate 2, frontispiece) have become available, the filaments have turned out to form dense sheets of galaxies, of which the largest one is the 'Great Wall' which extends across 170 h^{-1} Mpc length and 60 h^{-1} Mpc width [4]. The various superclusters emerge clearly from the Infrared Astronomical Satellite IRAS' survey of the whole sky out to distances of 140 h^{-1} Mpc [5] shown in Fig. 46. Confronting the CDM model with all the constraining data, it is clear that it fails. In Fig. 47 we have plotted the power function $P(k)$ normalized such that it fits the IRAS [8] two-point galaxy correlations for large values of k, or equivalently small angular and linear scales. The CDM curve then has too little power at large scales when compared to the COBE observations of CMB anisotropy [7]. The figure does not

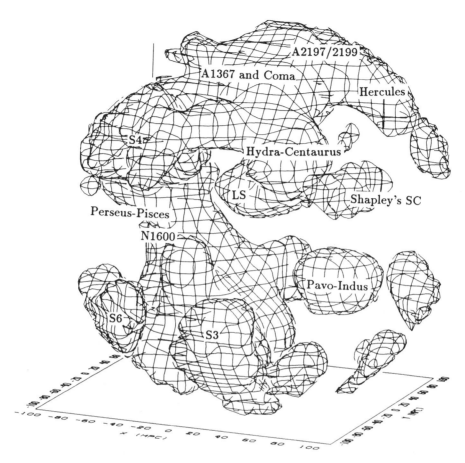

Fig. 46 The distribution of galaxies in the local universe. The plot shows isodensity contours drawn from an all-sky redshift survey of galaxies detected by the Infrared Astronomical Satellite. Galaxies in a sphere of radius 100 Mpc centred on the Milky Way are included—note that this represents at most 3% of the Hubble radius. The Galactic Centre points towards the right-hand side of the plot, along the positive *x*-axis with galactic longitude running anticlockwise around the sphere. The distribution has been smoothed with a Gaussian of dispersion 15 Mpc and the high-density regions enclosing roughly a third of the total volume are shown. Various well-known superclusters of galaxies are indicated. Adapted from the paper of Moore *et al.* [5] and provided by Dr C.S. Frenk (Durham University, UK)

show further constraints from bulk flows in the 15–40 Mpc/*h* range and X-ray clusters in the 20–70 Mpc/*h* range. Obviously, if the CDM power function had been normalized to fit the COBE large-scale data, it would have too much power on small scales. This situation is perhaps not so surprising since CDM was invented to explain structure on the scales of galaxies and small clusters. Isocurvature CDM models do not fare better, and in addition they predict an unrealistically short lifetime for the Universe, less than 9 Gyr.

What are then the possible amendments to the CDM model to make it tenable? So far we have considered the primordial seeds of the adiabatic CDM fluctuations to be

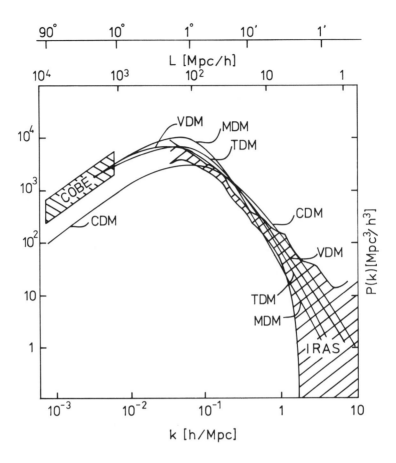

Fig. 47 Comparison of the power function *P(k)* of various dark matter models with the COBE measurements of radiation anisotropy and with the IRAS measurements of galaxy two-point correlations. The density fluctuation wave number *k* is given in units of *h*/Mpc. This can also be expressed by the angular scale in degrees, or by the linear size *L* of present galactic structures in units of Mpc/*h*. The normalization of the models is arbitrary. Here the plain CDM model is normalized to fit the IRAS data, whereas all the alternative models (described in the text) are normalized to fit the COBE data

Gaussian, leading to the Harrison–Zel'dovich spectrum with spectral index $n = 1$. This is, as we have noted in the previous section, in agreement with the COBE observations, but it is not strictly required. Better agreement with observations can be found if one supposes *ad hoc* that the initial density fluctuations were skewed towards larger wavelengths. These so called *tilted dark matter* models assume n to be smaller; in Fig. 47 the curve denoted TDM corresponds to $n = 0.7$. Clearly it is much more successful in fitting both the COBE and the IRAS data (as well as the intervening, not shown data).

An approach with a specific appeal is to assume the existence of topological defects remaining from the bubble walls of early phase transitions. Intersecting *domain walls* would give rise to *cosmic strings, monopoles,* and other textures, and intersecting strings would give rise to *cosmic loops*. These textures store enormous amounts of energy, so they could act as centres distorting the curvature, attracting WIMPs and causing non-Gaussian fluctuations. The density perturbations could then grow with a different time scale, not necessarily inflatoric, entering the non-linear regime earlier and thus forming galaxies earlier. This would result in smaller CMB anisotropy, and there would be more time to assemble the large-scale structures needed. Quantitatively, string theories cannot yet be tested, but they remain an interesting alternative to inflation and a field of intense speculation.

Another way out is to assume some very late phase transition, only after recombination. This would strongly perturb the mass distribution without affecting the already decoupled CMB, and thus anisotropy of the order of $\Delta T/T \simeq 10^{-6}$ could be reconciled with a very lumpy matter distribution on large scales.

One can also play with the dynamical parameters *ad hoc*. If Ω_0 were small, contrary to inflationary expectations, the sizes of mass structures for a given CMB angular scale would be larger. If the cosmological constant were different from zero, it might be possible to modify cosmic dynamics on large scales, leaving small scales alone. A positive λ reduces the deceleration, thereby increasing the age of the Universe for a given value of H_0. This fits inflation well since inflationary models really oversolve the flatness problem. Thus if the flatness condition were

$$\frac{\lambda}{3H_0^2} + \Omega_0 = 1$$

as in Eq. (3.86), a small Ω_0 and a large λ might produce a universe with many desirable features. The curve VDM in Fig. 47 represents such a vacuum-energy dark matter model, where CDM accounts for 20% of dark matter, and the cosmological constant accounts for the rest.

9.3 Hot Dark Matter

Neutrinos do not contribute to luminous matter, they are relativistic and their mass is not known, so they are HDM candidates. Their number density today from Eqs. (4.74) and (5.11) is

$$N_\nu = \frac{3}{11} N_\gamma \simeq 112 \, \frac{\text{neutrinos}}{\text{cm}^3} . \tag{9.22}$$

Thus the neutrinos, all species counted, could contribute a missing Ω_ν to close the Universe if their average mass were

$$m_\nu = \Omega_\nu \frac{\rho_c}{N_\nu} \simeq 95 \, \Omega_\nu \, h^2 \, \text{eV} . \tag{9.23}$$

We already know that the electron neutrino is too light to serve as dark matter, its mass being at most a few eV. As to the candidacy of the muon or tau neutrino, solar neutrino physics appears to rule out one of them. The flux of electron neutrinos reaching us from the Sun is too small compared with the calculable neutrino output in the Sun. An explanation consistent with all observations is that the electron neutrinos mix with the muon or tau neutrinos somewhere in the outer layers of the Sun and exchange identities, an effect called *resonant conversion*. This only works if the electron neutrino is very degenerate in mass with the mixing species. It follows that the latter also is too light to serve as a HDM candidate. We are then left with the third kind, for instance the τ neutrino, which is not required to mix with the electron neutrino. Although the τ neutrinos decoupled from the thermal plasma long before matter-domination they remained relativistic for a long time because of their small mass. This is the reason why they are called 'hot'.

Although the mass of the τ neutrino is poorly known and will probably remain so, cosmology itself dictates the value (9.23) in order to close the Universe. Equations (9.20) and (9.21) then show that free streaming τ neutrinos have such large velocities that they will only stop in overdensities of 40 Mpc linear size and 10^{15} $M_\odot$ mass. Thus the first structures formed are neutrino clouds of supergalactic size, all smaller fluctuations being erased by free streaming. This is then followed by the infall of baryonic matter after recombination.

In spherically symmetric clouds where the self-gravity is balanced by an isotropic thermal gas pressure, contraction in one direction is naturally compensated for by expansion in an orthogonal direction. Thus the first structures formed by contracting clouds are caustic surfaces best described by the *pancake model* of Zel'dovich (this applies to $k = 1$ closed universes). The distance between pancakes is about $40\Omega^{-1}h^{-2}$ Mpc, and they do resemble the large-scale sheets of galaxies. As the pancakes collapse, they disintegrate and form smaller structures. Thus HDM predicts a 'top-down' scenario where supergalaxies form first and galaxies later from the crumbs of the pancakes.

However, the 'top-down' scenario is contradicted by a number of observations. Supergalaxies are typically at distances $z \lesssim 0.5$, whereas the oldest objects known are the quasars with redshifts up to $z = 4 - 5$. There are also several examples of galaxies which are older than the groups in which they are now found. Moreover, in our neighbourhood the galaxies are generally falling in towards the Virgo cluster rather than streaming away from it, as one would expect in this scenario.

Computer simulations of pancake formation and collapse show that the matter at the end of the collapse is so shocked and so heated that they do not condense but remain ionized, unable to form galaxies and attract neutrino halos. Moreover, large clusters ($\lesssim 10^{14}M_\odot$) have higher escape velocities, so they should trap five times more neutrinos than large galaxies of size $10^{12}M_\odot$. This is not supported by observations which show that the ratio of dynamic mass to luminous mass is about the same in objects of all sizes.

An interesting quantal argument concerning neutrino clouds is due to Tremaine and Gunn [9]. In Section 4.3 it was briefly mentioned that two identical fermions cannot get too close to one another because of the Pauli exclusion force. By 'identical' is here meant that their momenta are degenerate to within Δp and they are in the same spin state. By 'too close' is meant a distance Δx, such that the product with Δp obeys Heisenberg's uncertainty relation

$$\Delta x \Delta p \lesssim \hbar . \tag{9.24}$$

As the neutrino cloud collapses this force builds up a degeneracy pressure which hinders the neutrinos from getting too densely packed. Only n_{spin} neutrinos fit into a phase space box of size

$$(\Delta x)^3 (\Delta p)^3 . \tag{9.25}$$

Let the radius of the cloud be R and the momentum of the neutrinos $m_\nu v$. Then the maximal allowed number of neutrinos is

$$N_{max} = n_{spin} \frac{(\Delta x)^3 (\Delta p)^3}{2\pi^3} \propto n_{spin} R^3 (m_\nu v)^3 . \qquad (9.26)$$

If the mass of the neutrino cloud is

$$M = m_\nu N < m_\nu N_{max} , \qquad (9.27)$$

one obtains a lower limit to the neutrino mass. For rich clusters of galaxies m_ν is $\gtrsim 5\sqrt{h}$ eV, and for galaxy haloes the neutrino mass is bounded by $30\sqrt{h}$ eV. However, for dwarf galaxies m_ν is required to exceed $150\sqrt{h}$ eV in conflict with the value (9.23) quoted for the mass required to close the Universe. In other words, light neutrinos cannot be responsible for the considerable amounts of dark matter present in dwarf galaxy haloes. Thus adiabatic HDM models are in trouble.

Isocurvature HDM models assume that the distribution of matter (scalar bosons, quarks and neutrinos) started out homogeneous, but that the baryonic matter clumped as a result of some mechanism, perhaps a spatially irregular graceful exit or an irregular phase transition. The dark matter is constituted by neutrinos which are still freely streaming at t_{eq}. The accretion of neutrinos to form haloes around the baryon clumps would be a much later process. The CMB is then very little perturbed by the clumps because most of the energy is in neutrinos and radiation.

Although this model is quite vague, it may be put to use if the tau neutrinos could be shown to have a mass in the right range to close the Universe. Unfortunately the possibility of an experimental verification is very remote, since a mass of a few tens of eV is only a millionth of the presently known limit. Isocurvature HDM models also have the drawback of predicting an unrealistically short age of the Universe, less than 8 Gyr.

Since no self-consistent set of simple initial conditions like CDM or HDM matches the observed Universe one may need a mixture of some kind of CDM to explain structures on small scales and HDM on large scales. A hybrid model of this kind assuming 30% neutrinos and 70% CDM, as suggested by Davis, Summers, and Schlegel [10] is drawn in Fig. 47 as the curve MDM. Clearly it is quite successful in fitting the data. Note that the contribution of MACHOs has not yet been taken into account in any of the power functions in Fig. 47.

It is also conceivable that the thermal history is more complicated than it has been described here. Late reionizations may have washed out the small-angle CMB fluctuations, destroying part of the CDM footprints but leaving the HDM predictions intact [11].

Problems

1. To derive Jeans wavelength λ_J and Jeans mass M_J, Eq. (9.15), let us argue as follows. A mass $M_J = \rho \lambda_J^3$ composed of a classical perfect gas will collapse gravitationally if its internal pressure $P = \rho kT/m$ cannot withstand the weight of a column of material

of unit area and height λ_J. Here m is the mean mass of the particles forming the gas. If we set the weight of the latter greater than or equal to P,

$$\frac{GM_J}{\lambda_J^2} \rho \lambda_J \gtrsim \frac{\rho kT}{m} \qquad (9.28)$$

we will obtain a constraint on the sizes of fragment which will separate gravitationally out of the general medium. Show that this leads to Eq. (9.15) [12].

2. Suppose that $h = 0.5$ and that the τ neutrinos have mass 20 eV. Discuss the properties of an additional heavy neutrino needed to close the Universe and to explain galaxy formation. Note that experiments require the heavy neutrino to have a mass greater than 45 GeV.

3. Assuming that galaxies form as soon as there is space for them, and that their mean radius is $30\,h^{-1}$ kpc and their present mean number density is $0.03h^3$ Mpc^{-3}, estimate the redshift at the time of their formation [2].

References

1. E. W. Kolb and M. S. Turner, *The Early Universe*, Addison-Wesley Publ. Co., Reading, Mass, 1990.
2. P. J. E. Peebles, *Principles of Physical Cosmology*. Princeton University Press, Princeton, New Jersey, 1993.
3. D. Lynden-Bell, S. M. Faber, D. Burstein *et al.*, *The Astrophysical Journal*, **326** (1988) 19.
4. M. J. Geller and J. P. Huchra, *Science*, **246** (1989) 897.
5. W. Moore, C. Frenk, D. Weinberg *et al.*, *Mon. Not. R. Astron. Soc.*, **256** (1992) 477.
6. Reported by J. Silk in *Nature*, **356** (1992) 741.
7. G. F. Smoot, C. L. Bennet, A. Kogut *et al.*, *Astrophysical Journal*, **396** (1992) L1.
8. W. Saunders, C. Frenk, M. Rowan-Robinson *et al.*, *Nature*, **349** (1991) 32.
9. S. Tremaine and J. E. Gunn, *Physical Review Letters*, **42** (1979) 40.
10. M. Davis, F. J. Summers and D. Schlegel, *Nature*, **359** (1993) 393.
11. R. L. Davis, H. M. Hodges, G. F. Smoot *et al.*, *Physical Review Letters*, **69** (1992) 1856.
12. F. H. Shu, *The Physical Universe*, University Science Books, Mill Valley, CA, 1982.

10 The Ultimate Questions

Although we now have pushed back as far as the Planck time, the question of the origin of the Universe remains unanswered. To go any further in the absence of a working theory is pure speculation. Similarly, to predict the longtime future of our Universe—leave alone other bubbles—also involves speculation. In this final chapter we shall do just that, to use whatever tools we have to imagine The Beginning and The End of our Universe.

In Section 10.1 we discuss how to avoid the intial singularity. Making use of the second law of thermodynamics we show that there is a difference between The Beginning and The End. This defines a thermo-dynamically preferred direction of time. It follows that oscillating universes are impossible, or they require as drastic assumptions as the initial singularity itself.

Since we do not know whether the Universe is open or closed, there are many possible scenarios for the ultimate future. In Section 10.2 we account qualitatively for the fate of some types of universes in the long term, following the discussion of Barrow and Tipler [1].

10.0 The Beginning

As we have seen in Section 3.2, Friedman's equations are singular $t = 0$. On the other hand, the exponential de Sitter solution (3.69), $S(t) = \exp(Ht)$, is regular at $t = 0$. In the Schwartzschild metric (3.7) the coefficient of dt^2 is singular at $r = 0$ whereas the coefficient of dr^2 is singular at $r = r_c$. However, if we make the transformation from the radial coordinate r to a new coordinate u defined by

$$u^2 = r - r_c,$$

the Schwartzschild metric becomes

$$d\tau^2 = \frac{u^2}{u^2 + r_c}dt^2 - 4(u^2 + r_c)du^2.$$

The coefficient of dt^2 is still singular at $u^2 = -r_c$ which corresponds to $r = 0$, but the coefficient of du^2 is now regular at $u^2 = 0$.

From these examples we see that some singularities may be just the consequence of a badly chosen metric and not a genuine property of the theory. Moreover, singularities usually exist only in exact mathematical descriptions of physical phenomena, but not in reality when one takes into account limitations imposed by observations. Or they do not exist at all, like the North Pole which is an inconspicuous location on a locally flat surface. Thus one might hope that these singularities in classical gravity also vanish under closer scrutiny.

However, according to theorems by Stephen Hawking and Roger Penrose they cannot be avoided because of the attractive nature of gravity. The classical metric is singular at $t = 0$ because time itself is created at that point. Since the proper distance between two events at x_i and $x_i + dx_i$ is given by the metric equation

$$ds^2 = g_{ik}dx_i dx_k ,$$

(10.1)

ds^2 vanishes when dx_i does. The only escape from this situation must come from considerations outside classical gravity such as quantum theory.

In quantum gravity the metric g_{ik} is a quantum variable which does not have a precise value, it has a range of possible values and a probability density function over that range. As a consequence, the proper distance from x_i to $x_i + dx_i$ is also a quantum variable which does not have a precise value; it can take on any value. It has a probability distribution which is peaked at the classical expectation value $\langle ds^2 \rangle$. Or, phrasing this a little more carefully, the quantum variable of proper distance is represented by a state which is a linear combination of all possible outcomes of an observation of that state. Under the extreme conditions near $t = 0$ nobody is there to observe, so 'observation' of $\langle ds^2 \rangle$ has to be defined as some kind of interaction occurring at this proper distance.

Similarly, the cosmic scale factor $S(t)$ in classical gravity has the exact limit

$$\lim_{t \to 0} S(t) = 0$$

(10.2)

at the singularity. In contrast, the quantum scale factor does not have a well-defined limit, it is fluctuating with a statistical distribution having variance $\langle S^2(t) \rangle$. This expectation value approaches a non-vanishing constant [2]

$$\lim_{t \to 0} \langle S^2(t) \rangle = C > 0 .$$

(10.3)

This resembles the situation of a proton–electron pair forming a hydrogen atom. Classically the electron would spiral inwards under the influence of the attractive Coulomb potential

$$V = \frac{e^2}{r} ,$$

(10.4)

which is singular at $r = 0$. In the quantum description of this system the electron has a lowest stable orbit corresponding to a minimum finite energy, so it never arrives at the singularity. Its radial position is random, different possible radii having some probability of occurring. Thus only the mean radius is well-defined, and it is given by the expectation value of the probability density function.

In fact, it follows from the limit (10.3) that there is a lower bound to the proper distance between two points,

$$\langle \mathrm{d}s^2 \rangle \geq \left(\frac{L_p}{2\pi} \right)^2 , \tag{10.5}$$

where L_p is the Planck length 10^{-35}m [2]. The light cone cannot then have a sharp tip at $t = 0$ as in the classical picture. Somehow the tip of the cone must be smeared out so that it avoids the singular point, forming a hemisphere.

James Hartle and Stephen Hawking have proposed a particular model of this kind. Their interpretation of the hemispheric modification of the tip of the light cone is that the time dimension gradually turns into a space dimension as one moves beyond Planck time. At the bottom of the hemisphere the four-dimensional spacetime has become a four-dimensional spatial manifold. Thus one can say that time emerges gradually from this space without any abrupt coming into being. Time is limited in the past, to within $\langle t^2 \rangle \simeq t_P^2$, but it has no sharp boundary. In that sense the classical singularity has disappeared, and there is no origin of time. As Hawking expresses it, 'the boundary condition of the Universe is that it has no boundary' [3].

The Universe then exists because of one unlikely fluctuation of the vacuum, in which the energy ΔE created could be so huge because the time Δt was small enough not to violate Heisenberg's uncertainty relation

$$\Delta t \, \Delta E \lesssim \hbar . \tag{10.6}$$

The Hartle–Hawking universe is, however, just one example of infinitely many solutions between which we cannot choose. Also, the step from this simple idea to a functioning theory of quantum gravity is very large and certainly not mastered.

Another approach to quantum gravity is the *conformal quantization* of J. Narlikar and collaborators [4]. Since quantization obviously spoils the structure of the light cone, one might look for non-trivial changes of spacetime which preserve the light cones as they were, in spite of quantization. Such transformations of the spacetime geometry exist. They imply the introduction of a conformal function Ω which is a function of spacetime, and which scales all lengths locally uniformly while preserving all the angles. In fact, the Robertson–Walker spacetime can be obtained from the Minkowski spacetime by a conformal transformation. The bonus is that quantization can be carried out, yet the light cone structure is preserved globally.

By this formal device Narlikar evicts the classically *certain* singularity, replacing it by a typically quantal *probable* singularity along with probabilities for other, non-singular origins of the Universe. Interestingly enough, it turns out that the probability for a singular origin is vanishingly small.

Another enigma is why spacetime was created four-dimensional. The classical answer was given by Paul Ehrenfest (1880–1933) who demonstrated that planetary systems with stable orbits in a central force field are possible only if the number of spatial coordinates is two or three. But today there are theories in which the most natural dimensionality is 11. Is the spacetime around us then accidentally four-dimensional or does it only appear so because seven more dimensions are curved at some extremely small *compactification scale*, perhaps the GUT scale ? Perhaps there are other universes like the bubbles in Linde's chaotic foam, where the dimensionality of spacetime in one bubble is unrelated to the next one.

From these ideas of a quantum fluctuating metric with non-vanishing expectation values for the minimum distance and time, many physicists have thought that a closed and oscillating universe would be a natural consequence. The present expansion would be followed by a contraction down to the minimum distance, bounce, and then a restart of the expansion of the next cycle. The origin of our present Universe would then just be the bounce at the end of the previous cycle of which we will never know anything. And since time is created along with each universe, there is no scale of time that would run continuously from one cycle to the next. Thus if the universe were carrying out such endless oscillations, it still does not mean that time is infinite.

That the oscillating universe entails more problems than it solves has been pointed out by Roger Penrose [2]. Recall the dictum of the second law of thermodynamics that the entropy of a system cannot decrease. In the radiation-dominated Universe the source of energy of photons and other particles is a phase transition or a particle decay or an annihilation reaction, many of these sources producing monoenergetic particles. Thus the energy spectrum at the source is very non-uniform and non-random, containing sharp spectral lines 'ordered' by the type of reaction. Such a spectrum corresponds to low entropy. Subsequently scattering collisions will redistribute the energy more randomly and ultimately degrade it to high-entropy heat. Thermal equilibrium is the state of maximum uniformity and highest entropy. The very fact that thermal equilibrium is achieved at some time tells us that the Universe must have originated in a state of low entropy.

In the transition from radiation domination to matter domination no entropy is lost. We have seen the crucial effect of photon reheating due to entropy conservation in the decoupling of the electrons. As the Universe expands and the wavelengths of the CMB photons grow, the available energy is continuously converted into lower grade heat, thus increasing entropy. This thermodynamic argument defines a *preferred direction of time*.

When the cooling matter starts to cluster and contract under gravity, a new phase starts. We have seen that the Friedman equations dictate instability, and the lesson from Chapter 9 is that the lumpiness of matter increases, with the Universe developing from an ordered, homogeneous state towards chaos. It may seem that the contracting gas clouds represent higher uniformity than matter clustered into stars and galaxies. If the only form of energy were thermal, this would indeed be so. It would then be highly improbable to find a gas cloud squeezed into a small volume if nothing hinders it from filling a large volume. However, the attractive nature of gravity seemingly reverses this logic: the cloud gains enormously in entropy by contracting. Thus the preferred direction of time as defined by the direction of increasing entropy is unchanged during the matter-dominated era.

The same trend continues in the evolution of stars. Young suns burn their fuel by fusion reactions in which energetic photons are liberated. During hydrogen burning the fusion reactions are the same as we have already described in connection with the primordial nucleosynthesis, Eqs. (5.23–5.35). Subsequently burning continues through a chain of fusion reactions in which heavier nuclei are always produced. The energy liberated mostly appears in the form of photons of MeV energies. As these photons diffuse in the stellar matter, they ultimately convert their energy into a large number of low-energy photons and heat, thereby increasing entropy.

Old suns may be extended low-density, high-entropy red giants or white dwarfs without enveloping matter. Heavy stars lose mass by various processes. If after all the loss a star ends up having a mass exceeding the *Chandrasekhar limit* $1.4\,M_\odot$, it ultimately undergoes a supernova explosion, collapsing into a neutron star which consists of highly uniform and degenerate neutron matter. The process is halted when the neutrons are as close to each other as is permitted by the Pauli exclusion force. In this process entropy grows enormously.

If the mass of the remnant exceeds the *Landau–Oppenheimer–Volkov limit* of about $2.5\,M_\odot$, the object finally collapses into a black hole. Matter in this form represents ultimate chaos and maximum entropy. For a black hole of horizon surface A, the entropy is proportional to A and given by the *Bekenstein–Hawking formula*

$$s = \frac{Akc^3}{4G\hbar} \ . \tag{10.7}$$

For a spherically symmetric black hole of mass m the surface area is given by

$$A = 8\pi m^2 G^2 / c^2 \ . \tag{10.7}$$

Inserting this into Eq. (10.7), entropy comes out proportional to m^2,

$$s = m^2 2\pi kGc/\hbar \ . \tag{10.9}$$

Thus two black holes coalesced into one possess more entropy than they both had individually.

Let us now consider the fate of a closed universe. After the time $t_{\max}$ given by Eq. (3.57) the expansion turns into contraction, and the density of matter grows. If the age of the universe is short enough that it contains black holes which have not evaporated, they will start to coalesce at an increasing rate. Thus entropy continues to increase, so that the preferred direction of time is unchanged. Shortly before the Big Crunch, when the horizon has shrunk to linear size L_p, all matter has probably been swallowed by one enormously massive black hole.

Therefore, since entropy has been increasing all the time, the Big Crunch singularity at $t = 2t_{\max}$ is of the same kind as the singularity in a black hole, and quite unlike the Big Bang 'white hole' singularity. Thus, the Universe cannot bounce and turn around to restart a new expansion cycle, the second law of thermodynamics forbids it. Speculations about a cycling universe are based on classical gravity and quantum mechanics, both of which are symmetric in the direction of time, but the second law of thermodynamics is not. The only way out would be to postulate a mechanism which would destroy the huge amount of entropy together with spacetime, letting the Universe tunnel out 'on the other side' with a virgin spacetime and minimal entropy, and preserving no memory of what has been before. However, that requires such drastic modifications of the laws of physics that the price is at least as high as to accept a mathematical singularity as the realistic end.

In his celebrated book *A Brief History of Time* [3], Stephen Hawking has discussed the direction of time which can be specified in three apparently different ways. Biological time is defined by the aging of living organisms like ourselves, and by our subjective sense of time, that we remember the past but not the future. But this can be shown to be a consequence of the thermodynamical direction of time which we have discussed above.

Apparently independent of this is the direction defined by the cosmic expansion which happens to coincide with the thermodynamical arrow of time. More recently Hawking, Laflamme and Lyons [5] have pointed out that the density perturbations also define a direction of time independent of the cosmic time, because they always grow, whether the Universe is in an expanding or a contracting phase.

The possible reason why the different arrows of time all point in the same directions is still obscure. Perhaps one has to resort to the *Anthropic Principle* [1] to 'understand' it, implying that the Universe could not be different if we should be able to exist and observe it. One would then conclude that the agreement between all the different directions of time speaks for the necessity of the Anthropic Principle.

10.2 The End

On a planetary scale, all biological life will end when the atmospheric carbon dioxide level has dropped below the minimum necessary for photosynthesis. According to recent estimates [6] this might happen in a mere 100 million years. This is much earlier than the time when the Sun is estimated to have spent its fuel, in about 5.5 Gyr. However, the cosmological time scale which interests us here is certainly much longer than that.

Given enough time, galaxies evolve by evaporating stars. Although they are bound systems of stars, there is a finite probability for stars to be accelerated above the escape velocity. These evaporated stars escape with positive energy, and thus the remainder of the galaxy is increasing its negative energy, as required by the law of energy conservation. Thus galaxies become more strongly bound with time, the end result being a black hole.

In open universes an important time scale is set by the proton lifetime—a speculative subject in itself. As we noted in Eq. (6.79), the proton lifetime τ_p is now estimated to be at least 23 orders of magnitude longer than the present age of the Universe, or 10^{24} Gyr. Current GUTs suggest proton decay to be caused by rare reactions turning protons into electrons as depicted in Fig. 41. It could be that the proton is indeed absolutely stable, so that its life is infinite. In that case ordinary matter in stars and galaxies will continue to exist until it is gobbled up by black holes, or perhaps forever.

Let us classify universes into shortlived ones with an existence less than τ_p, and longlived ones exceeding τ_p. Closed universes are found in both classes, but open or critical universes are always longlived.

Another time scale to take into account in longlived universes depends on whether monopoles exist [7]. As we have learned, monopoles have an onion-like structure with free X leptoquarks deep in their interior. The monopoles are so heavy that they should accumulate in the centre of stars where they may collide with protons. Some protons may occasionally penetrate in to the GUT shell and collide with a leptoquark, which transforms a d quark into a lepton according to the reaction

$$d + X_{virtual} \rightarrow e^+ \ . \tag{10.10}$$

This corresponds to the lower vertex in the drawing of the proton decay diagram, Fig. 41. Thus monopoles would destroy hadronic matter at a rate much higher than their natural decay rate. This would catalyze a faster disappearance of baryonic matter, thus yielding a different time scale for the Universe.

Yet another time scale is that of complete evaporation of black holes. From Eq. (3.16), black holes of solar size would evaporate within 10^{56} Gyr. The largest gravitationally bound objects today are superclusters with masses of the order of $10^{17} M_\odot$ or 2×10^{47} kg. Black holes of that size would last some 10^{106} Gyr. Clearly then, if ever the conditions permit black hole formation, they will probably outlast GUT protons by a large margin. If the matter density is very undercritical, black holes may never form, but since Ω_0 has been observationally found to be at least 0.1, they probably do form in our Universe, and we even believe we have found some.

A shortlived closed universe has a rather well predictable future [8]. As contraction proceeds, galaxies start to merge, and when the the ambient temperature becomes equal to the interior of the stars, they disintegrate by explosion. Stars are also disrupted by colossal tidal forces in the vicinity of black holes which grow by accreting stellar matter. As the temperature rises, we run the expansion history backwards to recover free electrons, protons and neutrons, subsequently free quarks, and finally free massless GUT fields at time t_{GUT} before the final singularity. The rôle of black holes is model-dependent, but it is reasonable to imagine that all matter ends up in one black hole. What then happens at the singularity escapes all prediction.

In a longlived universe proton decay is a source of heat for dead stars until about the time τ_p. More longlived universes, whether closed or open, will not contain any stars because they will have disappeared with the decaying protons. Thus matter will be composed of black holes, dark matter particles, monopoles and decay products of the protons: electrons, positrons, neutrinos and a considerable amount of radiation. These decay products are responsible for most of the heat and entropy because the relic CMB from the Big Bang has been redshifted away to completely negligible energies. The Universe is still matter-dominated if the neutrinos are massive, or in any case the dark matter particles necessary to close the Universe are.

If the longlived universe is closed it will be contracting very slowly. At some point the temperature will have risen high enough to permit mesons and baryons to reappear. Finally we join the picture of the shortlived closed universe as the expansion history is run backwards.

If the longlived universe is open, the future is boring [1]. Already long before time τ_p all stars have exhausted their fuel to become either black holes, neutron stars, black dwarfs or dead planets. After τ_p the radiation from decayed protons may cause a brief time of radiation-domination, of the order of 1000 τ_p. At later times the decay photons will have been redshifted to low enough energy to allow matter-domination to take over again. In the remaining leptonic (and dark non-baryonic) matter the only thing happening is the very slow formation of positronium by the free electrons and positrons. However, these atoms have little resemblance to present-day positronium. Their size will be of the order of 10^{15} Mpc, and they will rotate around their centre of gravity with an orbital velocity about 1 μm per century ! In the end, each of these atoms decay into some 10^{22} ultrasoft photons at a time scale comparable to the evaporation time of supergalaxy-sized black holes.

Another scenario is worth mentioning: a repulsive cosmological constant may, given enough time, convert even a closed universe into an ever expanding inflational de Sitter universe. Whatever matter remains when the exponential expansion takes over, it is hard to imagine that anything new would be formed thereafter, not even horizon-sized positronium atoms.

Our last speculation concerns the fate of black holes. Once a black hole has evaporated there will remain only a naked singularity, whatever that implies. The particles into which it evaporated have a perfectly thermal black-body spectrum, so they carry no information about the black hole. Then where did the huge amount of information or entropy in the black hole go ? Physics does not accept that it just vanished, so this problem has stimulated a flood of research.

Penrose and Hawking have conjectured that singularities should be protected from inspection, either because they exist entirely in the future (Big Crunch), or entirely in the past (Big Bang), or else they are hidden by an event horizon inside black holes. This is called the hypothesis of *cosmic censorship* which forbids the existence of naked singularities. Otherwise, since spacetime has its origin in a singularity, and spacetime ceases to exist in the singularity of a black hole, perhaps all of spacetime disappears when only naked singularities remain. In that case, even an open universe may turn out to have a finite lifetime.

Problems

1. Assume that the protons decay with a mean life of $\tau_p = 10^{35}$ yr, converting all their mass into heat. What would the ambient temperature on the surface of Earth be at time $t = \tau_p$, assuming that no other sources of heat contribute.

References

1. J. D. Barrow and F. J. Tipler, *The Anthropic Cosmological Principle*, Oxford University Press, Oxford, 1988.
2. R. Penrose, *The Emperor's New Mind*, Oxford University Press, Oxford, 1989.
3. S. W. Hawking, *A Brief History of Time*, Bantam Books, New York, 1988.
4. J. V. Narlikar, *The Primeval Universe*, Oxford University Press, Oxford, 1988.
5. S. W. Hawking, R. Laflamme and G. W. Lyons, *Physical Review*, **D47** (1993) 5342.
6. K. Caldeira and J. F. Kasting, *Nature*, **360** (1992) 721.
7. M. S. Turner, *Nature*, **306** (1983) 161.
8. M. J. Rees, *Observatory*, **89** (1969) 193.

Index